Genomics: Applications in Human Biology

GENOMICS

Applications in Human Biology

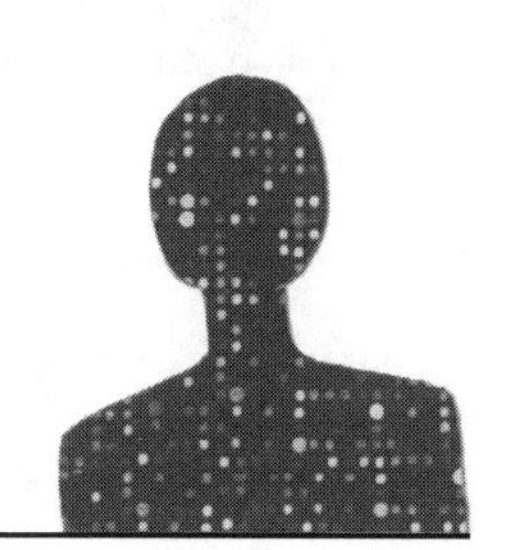

Sandy B. Primrose
Senior Partner, Business & Technology Management, High Wycombe, UK

Richard M. Twyman
Department of Biology, University of York, York, UK
Managing Director, Write Science, York, UK

BLACKWELL PUBLISHING
350 Main Street, Malden, MA 02148-5020, USA
9600 Garsington Road, Oxford OX4 2DQ, UK
550 Swanston Street, Carlton, Victoria 3053, Australia

First published 2004 by Blackwell Publishing Ltd

2 2005

Library of Congress Cataloging-in-Publication Data

Primrose, S. B.
Genomics : applications in human biology / Sandy B. Primrose and Richard Twyman.
p. ; cm.
Includes index.
ISBN 1-4051-0819-3 (pbk.)
1. Medical genetics. 2. Genomics. 3. Pharmaceutical biotechnology. 4. Molecular biology. I. Twyman, Richard M. II. Title.
[DNLM: 1. Genomics. 2. Biotechnology. 3. Molecular Biology.
QU 58 . 5 P953g 2004]
RB155 . P6936 2004
616′. 042—dc21

2003007541

ISBN-13: 978-1-4051-0819-5 (pbk.)

A catalogue record for this title is available from the British Library.

Set in 9½/12pt Photina
by Graphicraft Ltd, Hong Kong
Printed and bound in the United Kingdom
by TJ International Ltd, Padstow, Cornwall

The publisher's policy is to use permanent paper from mills that operate a sustainable forestry policy, and which has been manufactured from pulp processed using acid-free and elementary chlorine-free practices. Furthermore, the publisher ensures that the text paper and cover board used have met acceptable environmental accreditation standards.

For further information on
Blackwell Publishing, visit our website:
www.blackwellpublishing.com

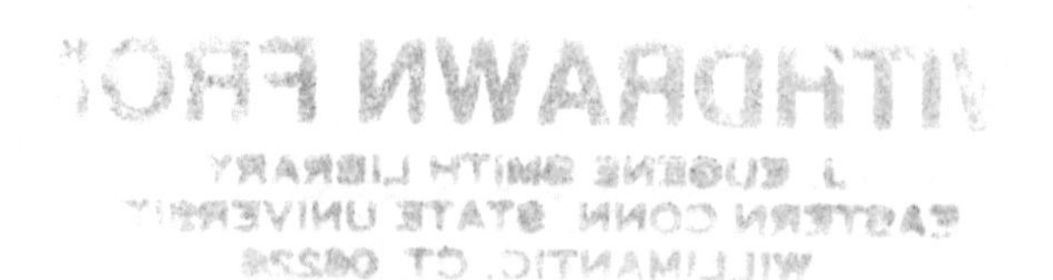

Brief Contents

Full Contents

CHAPTER SIX: The large scale production of biopharmaceuticals 131

CHAPTER SEVEN: Genomics and the development of new chemical entities 157

CHAPTER EIGHT: Gene and cell therapies 178

Preface

Fifty years ago, Watson and Crick detailed for us the structure of DNA and showed how it could be replicated faithfully from generation to generation. The impact of this discovery on medicine was barely considered. Rather, biologists wanted to know about the structure of genes and the genetic code. Twenty-five years ago the biotechnology revolution was underway following the development of recombinant DNA technology, which permitted the *in vitro* production of human proteins on a large scale. Then the vision for biotechnology was no more than factories producing recombinant molecules. Pharmaceutical biotechnology, as it then was known, was a very narrow subject.

Today we are in the midst of the genomics revolution, which was spearheaded by international projects aiming to sequence the complete genomes of organisms ranging from bacteria to mammals, including humans. Many of the genes in these organisms have been identified and good progress is being made towards understanding the roles of these genes in health and disease. As a consequence, there is almost no aspect of medicine and drug development that has not been affected. For example, we now have a good understanding of the genes involved in microbial pathogenicity and this is facilitating the development of new diagnostics, new vaccines, and new antibiotics. Similarly, we are rapidly dissecting the genetic basis of inherited diseases and cancer, which again is leading to new diagnostics and new treatments. The development of these new pharmaceuticals is being facilitated by the introduction of novel screening methodologies that are themselves based on recombinant DNA technology and genomics.

When Watson and Crick announced their momentous discovery almost all pharmaceuticals were small molecules, although insulin was a notable exception. Following the advent of recombinant DNA technology this drug repertoire was expanded to include a much wider range of natural human proteins including interferons, blood products, and further hormones. Today the diversity of drug molecules has expanded further, to include engineered proteins that are unlike any produced naturally, humanized antibodies, and even nucleic acids. Furthermore new medical procedures are being developed, such as gene therapy, cell therapy, and tissue therapy.

Given the pace at which the above developments are taking place it is not surprising that students and their academic mentors have difficulty in seeing the whole picture. This book has been written to provide them with the necessary overview, covering technologic developments, applications, and (where necessary) the ethical implications. The book is divided into three sections. The first section (Chapters 1 and 2) introduces the role of biotechnology and genomics in medicine and sets out some of the technologic advances that have been the basis of recent medical breakthroughs. The second section (Chapters 3–5) takes a closer look at how biotechnology and genomics are influencing the prevention and treatment of different categories of disease. Finally, in the third section (Chapters 6–8), we describe the contribution of biotechnology and genomics to the development of different types of therapy, including conventional drugs, recombinant proteins, and gene/cell therapies.

Throughout the book, the level of detail has been selected so that the reader can grasp what has been achieved without falling victim to "not seeing the wood for the trees." A basic understanding of genetics and molecular biology has been assumed so we can avoid the obligatory chapters on DNA structure, gene expression, etc. that appear in most larger biology textbooks regardless of their actual focus. Readers requiring more detail of the recombinant DNA and genomics techniques should consult our more advanced textbooks on these subjects: *Principles of Gene Manipulation* (POGM) and *Principles of Genome Analysis and Genomics* (POGA), also published by Blackwell Publishing. References to appropriate sections in these two books are included at the end of each chapter (with the relevant acronym indicating the book), plus a short bibliography mostly comprising review papers that have been selected for their clarity of presentation. The reader will also find the text contains several categories of boxed text, which include history boxes (describing the origins and development of particular technologies or treatments), molecular boxes (which describe the molecular basis of diseases or treatments in more detail), and ethics boxes (which discuss the ethical implications of technology development and new therapies).

Finally, we would like to thank the people who provided invaluable assistance in the preparation of the manuscript, particularly Sue Goddard and her team in the library at CAMR and Alistair Fitter at the Department of Biology, University of York. Richard Twyman would like to dedicate this book to his parents, Peter and Irene, his children, Emily and Lucy, and to Hannah, Joshua, and Dylan.

Sandy B. Primrose and Richard M. Twyman

References

Primrose SB, Twyman RM (2003) *Principles of Genome Analysis and Genomics*, 3rd edn. Blackwell Publishing, Oxford.

Primrose SB, Twyman RM, Old RW (2001) *Principles of Gene Manipulation*, 6th edn. Blackwell Science, Oxford.

Acknowledgments

Some figures and tables have been used from other sources. We thank the various authors and publishers for permission to use this material, which has come from the following sources:

Figures are extensively drawn from the following publications by the authors:

Primrose SB (1991) *Molecular Biotechnology*, 2nd edn. Blackwell Science, Oxford.
Primrose SB, Twyman RM (2003) *Principles of Genome Analysis and Genomics*, 3rd edn. Blackwell Publishing, Oxford.
Primrose SB, Twyman RM, Old RW (2001) *Principles of Gene Manipulation*, 6th edn. Blackwell Science, Oxford.

Specific tables and figures have been taken from the following sources:

Fig. 2.4: Coulson A, Sulston J, Brenner S *et al.* (1986) Toward a physical map of the genome of the nematode *Caenorhabditis elegans*. *Proc Natl Acad Sci USA* **83**, 7821–7825.
Fig. 2.8: EnsEMBL human genome browser www.ensembl.org
Fig. 2.9: Veculescu VE *et al.* (1997) Characterization of the yeast transcriptome. *Cell* **88**, 243–251.
Fig. 2.12 inset: Görg A, Postel W, Baumer M, Weiss W (1992) Two-dimensional polyacrylamide gel electrophoresis, with immobilized pH gradients in the first dimension, of barley seed proteins: discrimination of cultivars with different mating grades. *Electrophoresis* **13**, 192–203.
Fig. 3.4: Courtesy of Catherine Arnold, UK Health Protection Agency.
Fig. B3.3: Behr *et al.* (1999) *Science* **284**, 1520–1523. [for Box 3.3]
Fig. 4.4: Nussbaum RL, McInnes RR, Willard HF (2001) *Genetics in Medicine*, WB Saunders, Philadelphia, figure 4.14. Original photograph courtesy of P. Wray, Hospital for Sick Children, Toronto.
Fig. 4.6: Nussbaum RL, McInnes RR, Willard HF (2001) *Genetics in Medicine*, WB Saunders, Philadelphia.

Fig. 4.7: Thomson G (2001) Mapping of disease loci. In: Kalow W, Meyer UA, Tyndale R, eds. *Pharmacogenomics*, pp 337–361. Marcel Dekker, New York.
Fig. 4.9: Judson R, Stephens JC, Windemuth A (2000) The predictive power of haplotypes in clinical response. *Pharmacogenomics* **1**, 15–26.
Fig. 4.10: Nussbaum RL, McInnes RR, Willard HF (2001) *Genetics in Medicine*, WB Saunders, Philadelphia, figure 4.13.
Fig. 4.11: Johnson JA, Evans WE (2002) Molecular diagnostics as a predictive tool: genetics of drug efficacy and toxicity. *Trends Mol Med* **8**, 300–305.
Fig. 5.6: Funaro A, Hovenstein AL, Santoro P *et al.* (2000) Monoclonal antibodies and therapy of human cancers. *Biotechnol Adv* **18**, 385–401, figure 2.
Fig. B6.4b: Procognia Ltd.
Fig. 7.4: Croston GE (2002) Functional cell-based uHTS in chemical genomic drug discovery. *Trends Biotechnol* **20**, 110–115, figure 2.
Fig. 7.5: Bandara, Kennedy (2002) *Drug Discovery Today* **7**, 411–418, figure 2.
Fig. 7.7: Thompson, Ellman (1996) *Chem Rev* **96**, 555, figure 10.29.
Fig. 7.8: Balkenhol F, von dem Bussche-Hunnefeld C, Lansky A *et al.* (1996) *Angew Chem Int Ed Engl* **35**, 2289, figure 10.30.
Fig. 7.12: Castle AL, Carver MP, Mendrick DL (2002) Toxicogenomics: a new revolution in drug safety. *Drug Discovery Today* **7**, 728–736, figure 4a.

Table 7.1: Croston GE (2002) Functional cell-based uHTS in chemical genomic drug discovery. *Trends Biotechnol* **20**, 110–115.
Table 7.2: DeVito JA *et al.* (2002) An array of target-specific screening strains for antibacterial discovery. *Nature Biotechnol* **20**, 478–483.

CHAPTER ONE

Biotechnology and genomics in medicine

Introduction

Over the last 300 years, there has been a growing understanding of how the human body functions in health and disease. However, our knowledge has not increased steadily. The history of medicine is punctuated by sudden breakthroughs and leaps of innovation. Very few of these key developments would have been possible without underlying advances in **technology**.

As an example, consider the discovery of the first two antimicrobial substances by Alexander Fleming – lysozyme in 1922 and penicillin in 1928. Both discoveries were serendipitous, and neither would have been made if Fleming had been unable to culture bacteria on a solid growth medium. The use of agar for this purpose, initially proposed by Fanny Hesse, was put into practice by Robert Koch in 1882. Armed with such pure culture techniques, Robert Koch and Louis Pasteur were able to establish the principles of bacterial pathogenicity, thus founding the modern discipline of medical microbiology. In turn, the work of Fleming, Pasteur, and Koch stemmed from the discovery of bacteria by Anton van Leeuwenhoek in 1683, and this would have been impossible without the microscope. Van Leeuwenhoek made his own crude microscopes, but credit for the original invention goes to Hans and Zacharias Janssen in 1595. Similarly, the use of ether as an anesthetic, first demonstrated by Crawford Long in 1842,* would not have been possible without a method for ether synthesis. Such a method was first described by the German scientist Valerius Cordus in 1540. Thus, medical breakthroughs invariably have depended on technologic advances in physics, chemistry, and biology.

Since 1970, we have witnessed an unprecedented number of new medical innovations reflecting our increasing knowledge of the **molecular** basis of health and disease. While chemistry and physics have played their roles, much of this innovation is the direct result of two technologic revolutions in biology – the

* Crawford Long was the first to demonstrate the use of ether as an anesthetic, but provenance is often attributed to William Morton, who was the first to publish on the technique, in 1846.

recombinant DNA revolution and the **genomics revolution**, which are the subjects of this book. In this first chapter, we briefly summarize the impact of recombinant DNA and genomics on the practice of medicine. In later chapters, we discuss the role of these technologies in the prevention, diagnosis and treatment of different types of disease, and examine the emerging technologies that may contribute to the medical breakthroughs of the future.

Recombinant DNA technology

The **recombinant DNA revolution** began in about 1972 with the development of tools and techniques for *in vitro* DNA manipulation. Until the 1970s, it was impossible to manipulate DNA precisely, which meant it was very difficult to study individual genes in a direct manner. In model organisms, genetic analysis could be used to find out about the structure and function of genes indirectly, but such methods could not be applied easily to humans. Recombinant DNA technology was enabled by the isolation and biochemical characterization of enzymes that bacteria use to manipulate DNA as part of their normal cellular processes (Box 1.1). It was soon realized that if such enzymes could be purified, they could be used to create novel combinations of different DNA fragments *in vitro*. Such novel fragments were termed **recombinant DNA** molecules.

The central importance of cloning

To study a particular DNA sequence experimentally it is necessary to generate enough copies for laboratory-scale handling. The first significant advance offered by recombinant DNA technology was the ability to prepare millions of copies of the same DNA sequence, a technique known as **molecular cloning**. Researchers had

Box 1.1 Key enzymes used to manipulate DNA

- Restriction endonucleases. These are bacterial enzymes that cut DNA molecules internally at positions defined by specific target sequences, allowing large DNA molecules to be cut into predictable fragments. Both DNA strands are cut and the cleavage sites may be opposite each other (generating blunt fragments) or staggered (generating overhangs).
- DNA ligases. These are enzymes that join DNA fragments end to end. Some can join blunt fragments, while others require overhangs. The compatibility of overhanging ends depends on the restriction endonuclease used.
- DNA polymerases. These are enzymes that synthesize DNA on a complementary template. Different enzymes are used for DNA labeling, DNA sequencing, the polymerase chain reaction, and reverse transcription of mRNA into cDNA.
- DNA modification enzymes. Examples include alkaline phosphatase (which removes phosphate groups from the ends of DNA fragments) and polynucleotide kinase (which carries out the reverse process). These enzymes are used to control ligation reactions and for DNA labeling.

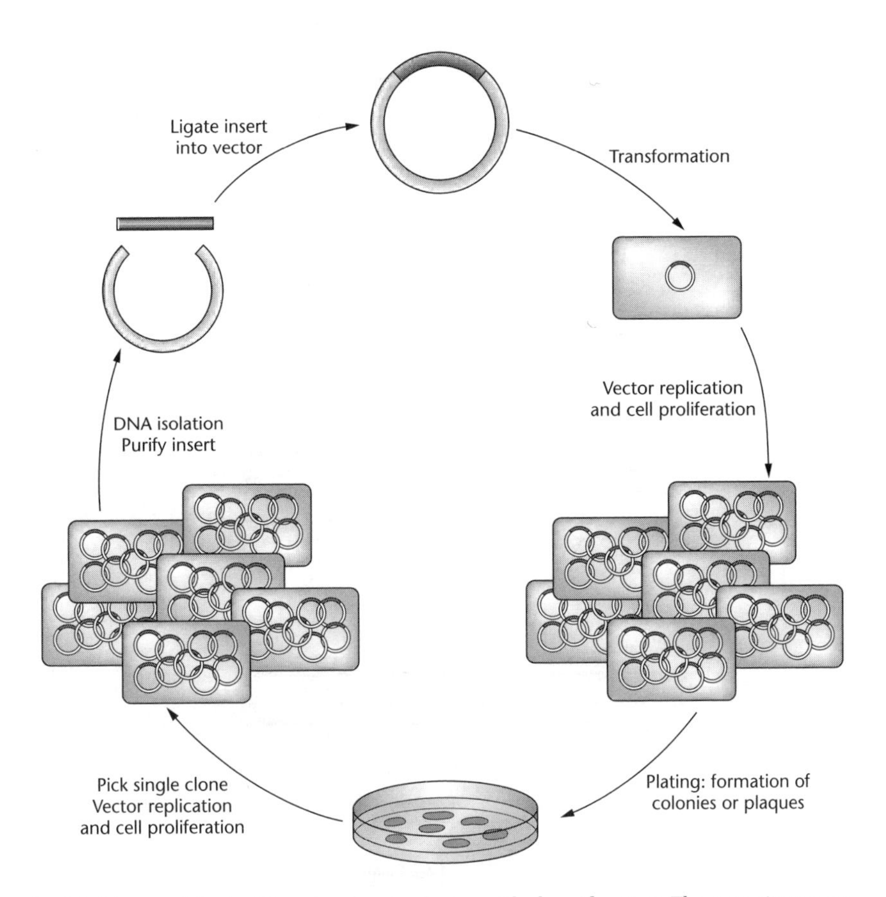

Fig. 1.1 The principle of cell-based molecular cloning with plasmid vectors. The vector is cut open with a restriction enzyme that has only one recognition site in the vector sequence, thus cutting it at a predictable position. The insert, prepared with the same enzyme, is sealed into place with DNA ligase. The recombinant vector is then introduced into the bacterium *Escherichia coli* by transformation. The vector carries a selectable marker gene (see p. 184) which allows transformed bacteria, but not normal bacteria, to survive and proliferate. When the bacteria are spread on a plate of medium supplemented with antibiotic, transformed bacteria form colonies containing about 1×10^6 cells in which each cell carries several hundred copies of the plasmid. Individual colonies are picked and grown in larger scale culture vessels under selection from which large amounts of DNA can be isolated. The insert, now massively amplified, can be purified using the same restriction enzyme used to insert it into the vector in the first place.

known for a long time that bacteria contained **autonomous replicons**, i.e. genetic elements such as plasmids and bacteriophage (phage) with the intrinsic ability to replicate to a high copy number. Recombinant DNA techniques were used to join such replicons to human DNA sequences, so that the human sequences were **amplified**. This principle led to the development of **cloning vectors**, i.e. DNA elements based on plasmids, phage, or sometimes a combination of both, which are used specifically to clone fragments of donor or passenger DNA. The general technique for cell-based molecular cloning is shown in Fig. 1.1.

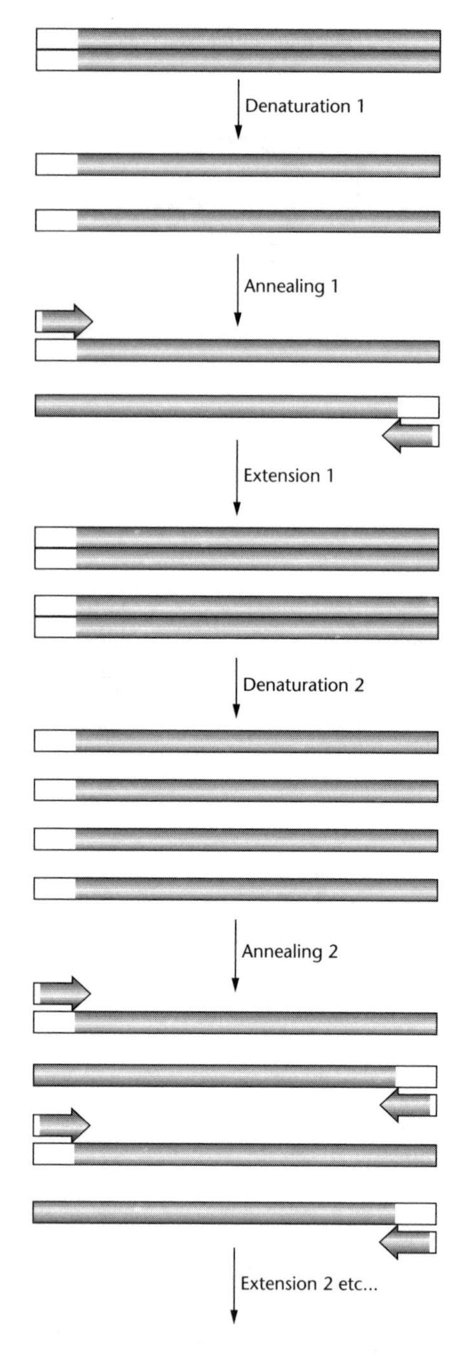

Fig. 1.2 The basic polymerase chain reaction. A double-stranded DNA template is denatured (separated into single strands) and two primers are annealed. The primers face towards each other, anneal to opposite strands, and define the target fragment to be amplified. Primer extension copies the DNA in the region between the two primers and therefore doubles the amount of template. The process of template denaturation, primer annealing, and primer extension is repeated 25–30 times. In the presence of excess primers and other reaction components, 25 cycles can theoretically yield over 8 million copies of the same fragment.

Box 1.2 Gel electrophoresis

Gel electrophoresis is the standard method for the size-separation of mixtures of DNA molecules. The basic principle is that DNA molecules in solution are negatively charged, and will therefore move towards the anode in an electric field. If the solution is dispersed within a matrix such as an agarose or polyacrylamide gel, the pores of the gel have a sieving effect, so that smaller molecules move towards the anode more rapidly than larger ones. The separating range of the gel depends on the pore size, which depends on the gel concentration. For example, a 5% agarose gel will separate DNA molecules within the range 100–500 bp, while a 0.5% gel will separate molecules in the range 5–20 kb. Polyacrylamide gels are used for smaller DNA fragments, and where it is necessary to distinguish between molecules differing in size by a single nucleotide (e.g. in DNA sequencing). In agarose gels, the fate of individual DNA molecules is followed using the intercalating fluorescent dye ethidium bromide, whereas in polyacrylamide gels the DNA is generally labeled prior to separation. Special techniques, such as pulsed-field gel electrophoresis, are required to separate molecules greater than 50 kb.

In the mid-1980s, a different technique for DNA amplification was developed that is carried out *in vitro* using purified DNA polymerase. This has become known as the **polymerase chain reaction (PCR)**. The basic PCR is shown in Fig. 1.2. The technique requires **primers**, single-stranded DNA molecules that anneal at particular sites on the template DNA. If two primers are designed to flank a target region of interest, face inwards, and anneal to opposite DNA strands, DNA synthesis across the region defined by the primers will double the amount of template available. Therefore, cyclical rounds of **denaturation** (separation of the template DNA into single strands), **primer annealing**, and **primer extension** by DNA synthesis can result in the exponential amplification of the target DNA sequence. Compared to traditional cell-based DNA cloning, the PCR is rapid, sensitive, and robust. It can be used to prepare large amounts of a specific fragment starting from a very small amounts of starting material, and that starting material does not have to be well preserved. For example, DNA can be extracted and amplified from fixed biologic specimens, blood and semen samples at crime scenes, and even Neanderthal bones! However, the PCR is generally less accurate than cell-based cloning because the DNA polymerases used in this procedure are error-prone. The standard technique is suitable for the amplification of fragments only up to about 5 kb in length, whereas large-capacity cloning vectors can easily amplify sequences that are several hundred kilobases long. Therefore cell-based cloning and the PCR have complementary although overlapping uses in human molecular biology.

Both of the cloning methods discussed above require a procedure that allows the progress of reactions to be followed and the products to be analyzed. The standard technique is **gel electrophoresis**, which separates DNA molecules on the basis of size (Box 1.2).

Identification and cloning of specific genes

Before a specific gene sequence can be cloned, it must be isolated from its natural source, and this is generally the bottleneck in any cloning procedure. The two

Table 1.1 Properties of genomic DNA and cDNA.

Genomic DNA	cDNA
With rare exceptions, genomic DNA is the same in all tissues from the same organism	cDNA differs between tissues, and according to developmental stage and cell state
Genes in natural context (includes spacer DNA, regulatory elements, and introns)	Only transcribed sequences represented. No spacer DNA, regulatory elements, or introns. Splice variants represented by different cDNAs
All genes represented	Only genes expressed in the tissue from which mRNA was obtained are represented
Genes represented equally	Different genes are not represented equally – strongly expressed genes will produce more transcripts and give rise to more cDNA copies than weakly expressed genes

major sources of DNA for cloning, genomic DNA and complementary DNA (cDNA), are both incredibly complex (Table 1.1). Individual genes are therefore diluted by millions of irrelevant DNA fragments.

In some rare cases, obtaining the desired sequence has been relatively straightforward. For example, among the first human genes to be cloned were those encoding α-globin and β-globin because the mRNA is so highly enriched in reticulocytes (immature red blood cells) that cDNA clones could be obtained simply by random sequencing. However, few genes fall into this "superabundant" category and more sophisticated strategies are usually required.

In cell-based molecular cloning, the general approach is to create a **DNA library**, in which a collection of cloned DNA fragments is assembled representing the entire source population (genomic DNA or cDNA). The library is then **screened** using one of the following procedures:

• Sequence-dependent screening. This is performed either by hybridization, using a labeled DNA or RNA probe (Box 1.3), or by PCR. In each case, the technique relies on the probe or PCR primer combination recognizing a particular clone in the library because it has the complementary sequence. Suitable probes or primer combinations can be obtained from existing partial clones, from clones of similar genes in other species, from consensus sequences representing a particular gene family, or from the known amino acid sequences of proteins.

• Immunologic screening. This requires an **expression library**, i.e. a cDNA library in which all the clones are expressed to produce proteins. If an antibody is available that recognizes the protein product of the target gene, the corresponding DNA clone can be isolated.

• Functional screening. This also requires an expression library. The screening procedure is a test for protein function, e.g. a particular enzyme activity or a particular effect when introduced into cultured cells.

In contrast to cell based cloning, the PCR can be used to isolate DNA sequences directly from the source (i.e. without first creating a library), essentially following a sequence-dependent screening strategy. As stated above, the standard PCR can

Box 1.3 Nucleic acid probes and hybridization

Hybridization, i.e. complementary base pairing between single-stranded nucleic acids, is one of the core techniques in molecular biology. It allows the identification of specific DNA sequences in complex mixtures. One nucleic acid molecule is **labeled** in some way to facilitate detection and then used as a **probe** to identify a specific target. For example, in **Southern blot hybridization**, genomic DNA is fragmented, separated by agarose gel electrophoresis, and then transferred to a membrane where it is immobilized as an imprint of the gel. The DNA is then denatured (to separate the strands) and a probe is added. The probe will hybridize to a specific target and will be revealed as a band when the label is detected (Fig. B1.3). Analogous procedures can be used to identify specific RNA molecules in mixtures separated by electrophoresis (**northern blot hybridization**) or RNA molecules *in situ* in tissue sections, embryos, or explants (***in situ* hybridization**). Hybridization is also used to identify clones in library screens (**colony or plaque hybridization**).

Traditionally, DNA and RNA probes have been labeled with radioactive substrates and detected by **autoradiography** (exposure to a radiation-sensitive film) or **phosphorimaging** (exposure to a radiation-sensitive screen). However, radioactive labels are being progressively replaced by nonradioactive alternatives, such as fluorophores, enzymes that can be detected using a colorimetric assay, chemiluminescent substrates, and haptens (which are detected with antibodies). Whatever label is used, incorporation involves either DNA/RNA synthesis with labeled nucleotide analogs or end-labeling reactions using DNA modification enzymes (Box 1.1).

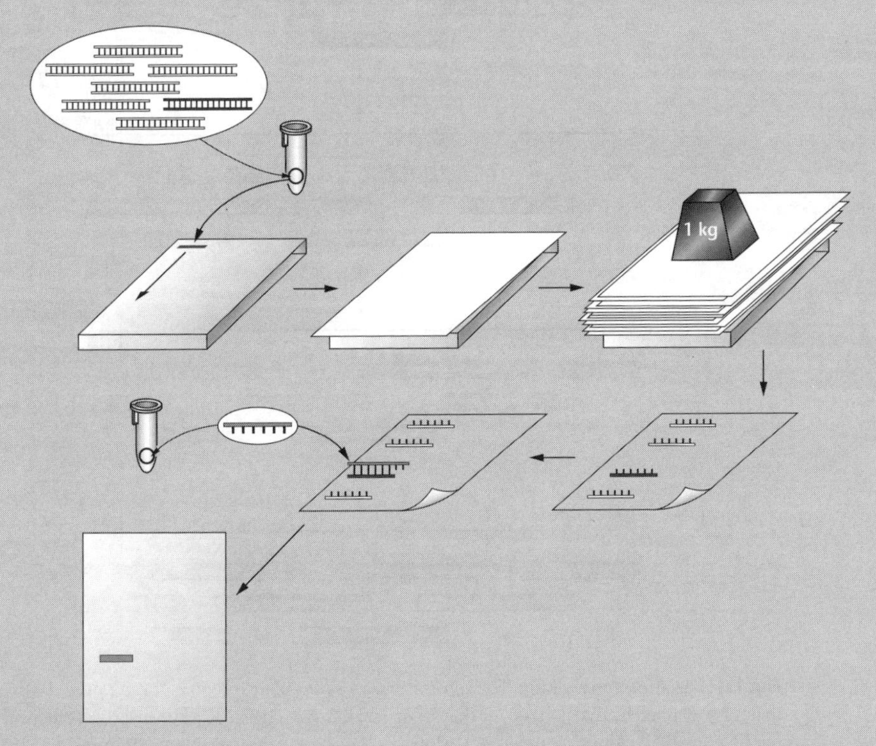

Fig. B1.3 The Southern blot demonstrates the value of hybridization in molecular biology. A complex population of DNA molecules (e.g. cDNA, digested genomic DNA) containing a target sequence of interest (shown in bold) is separated by electrophoresis and transferred onto a membrane by capillary blotting. This involves placing the membrane on top of the gel and then stacking absorbent paper on top, so that the buffer is drawn through and the DNA is transferred at the same time. The buffer is usually alkali so that the DNA is denatured into single strands at the same time. The immobilized DNA is then hybridized with a labeled probe recognizing the target. When the signal is detected, a single band is revealed on the membrane.

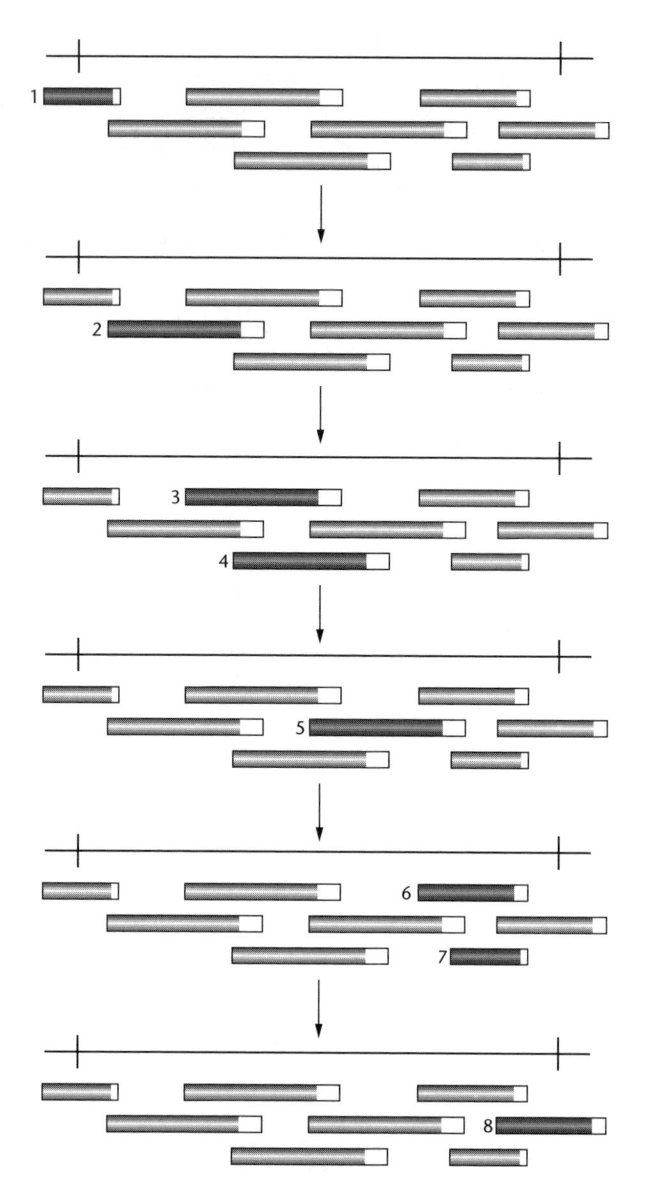

Fig. 1.3 Chromosome walking. The top line shows a candidate region of the genome, 1 Mb in length, defined by two genetic markers (vertical lines). Underneath, the inserts of different overlapping BAC clones are arranged to form a clone contig map. To create this map, one of the genetic markers (e.g. a restriction fragment length polymorphism (RFLP) or a microsatellite) is used as a probe to screen a BAC library, identifying clone 1. If the end of clone 1 is used as a probe, clone 2 is identified. Similarly, clone 2 will identify clones 3 and 4, either of which will find clone 5. Finally, clone 5 will hybridize to clones 6 and 7, either of which will identify clone 8. Clone 8 will also hybridize to the second genetic marker, therefore generating a bridge of clones spanning the candidate interval.

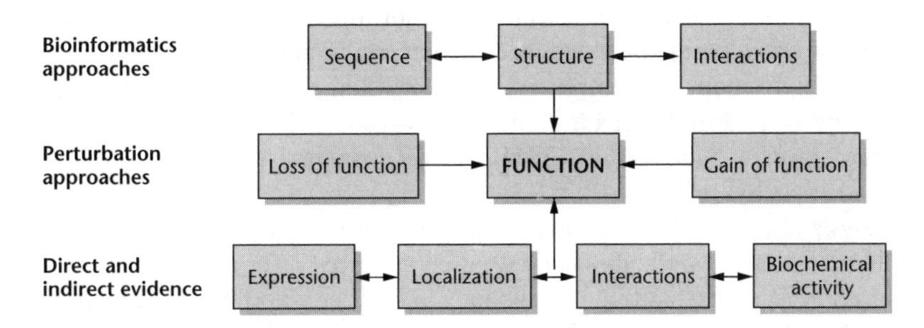

Fig. 1.4 A selection of approaches to study gene function on a global scale. Computers can be used to analyze protein sequences and structures, and predict their interactions from structural data, providing tentative functional annotations on the basis of information from related sequences and structures. Functions can be identified directly by mutation or interference to cause loss of function or by overexpression/ectopic expression to cause gain of function. Further evidence can be derived from mRNA/protein expression experiments, protein localization, direct experimental investigation of protein interactions, and assays for biochemical activity. These approaches are described in more detail in Chapter 2.

amplify fragments up to about 5 kb in length. However, the more recent innovation of **long PCR**, which employs a mixture of DNA polymerases, can amplify much larger fragments (up to 50 kb). **Reverse-transcriptase PCR (RT-PCR)** is the standard procedure for amplifying cDNA directly from a source of mRNA. The RT-PCR is a single-tube reaction where mRNA is first reverse transcribed and the cDNA is then amplified.

The above methods can be applied only if a suitable probe/primer combination can be designed or if some functional information is available about the target gene. This is not the case for most human disease genes because generally the only information available is the overall disease phenotype. A widely used approach under these circumstances is **positional cloning**, where the disease gene is first mapped genetically to a particular genomic region. Known DNA sequences in the vicinity, generally the genetic markers used for the initial mapping study but sometimes other landmarks such as chromosome breakpoints, are then used to initiate a **chromosome walk** in which overlapping genomic clones are identified by library screening until the candidate interval is covered (Fig. 1.3). This interval is then searched for genes, with the ultimate aim of finding a gene that carries a mutation in individuals suffering from the disease but not in healthy individuals.

Functional characterization of cloned genes

The cloning of a gene, e.g. a human disease gene, is only the first step in a long process. Once a clone is available, it is important to learn as much about the gene as possible, since this provides an insight into its normal function in the cell and its role in disease pathogenesis. A thorough understanding of the function of a gene in health and disease is valuable in the development of new therapies. There are many ways to learn about gene function (Fig. 1.4):

• Analysis of gene expression. Gene expression may be restricted to particular cells or tissues, to particular stages of development, or may be induced by external signals (e.g. hormones). Changes in gene expression patterns may be relevant in pathogenesis, and mutations in one gene may affect the expression patterns of others. Gene expression can be studied by methods such as northern blot hybridization and *in situ* hybridization (Box 1.3).
• Analysis of protein localization. If the gene can be expressed to produce a recombinant protein, antibodies can be raised and used as probes to study protein localization. Western blotting is analogous to northern blotting, and involves the separation of protein mixtures by electrophoresis followed by the use of antibody probes to detect specific proteins. Precise localization patterns in tissues and even within cells can be determined by *in situ* immunochemical analysis.
• Analysis of protein interactions. A number of genetic and biochemical techniques can be used to investigate protein interactions with other proteins, with nucleic acids, and with small molecules. This can help to determine gene function at the molecular and cellular levels and can link proteins into complexes or pathways.
• Altering gene expression or activity. Once a gene has been cloned, strategies can be developed to deliberately mutate that gene or to eliminate its function by interfering with its expression or the activity of its product. There are many different techniques that can be applied to study **loss of gene function**, including random mutagenesis, targeted gene mutation, interference with gene expression using antisense RNA, ribozymes or RNA interference, and interference with protein activity using antibodies (see Chapter 8). Conversely, the overexpression of a gene, expression outside its normal spatial or temporal domain (ectopic expression), or the expression of a mutant version of the protein that is more active than normal can be used to determine the consequences of **gain of gene function**. Such techniques can help to elucidate gene function at the cellular and whole organism levels, and can be used to create models of human diseases in cells and animals.
• Analysis of protein structure. If the structure of the encoded protein is solved, interactions with other proteins and small molecules can be modeled.

From recombinant DNA to molecular medicine

The initial medical advances made possible by recombinant DNA technology reflected the isolation and characterization of individual genes with medical relevance, i.e. human disease genes, related genes from other animals, and genes from pathogenic organisms. As well as increasing our fundamental knowledge of the molecular basis of human diseases, this allowed the development of a new field of medicine, termed **molecular medicine**, which is the direct application of recombinant DNA techniques to the prevention, diagnosis and treatment of human disease. A whole new biotechnology industry has grown up around the potential of molecular medicine and several key areas are discussed below.

The use of DNA sequences as diagnostic tools

One of the first direct medical applications of recombinant DNA technology was the use of DNA sequences as diagnostic tools. In the same way that probes or PCR primers can be used to isolate genes from clone libraries, they can also be used to detect DNA sequences related to disease. Importantly, no disease symptoms need to be evident. For example, inherited disorders can be detected prenatally (e.g. by chorionic villus sampling) or before the onset of symptoms (in the case of a late-onset diseases like Huntington's disease). Similarly, hybridization-based tests or PCR assays can be used to detect pathogens or malignant cells before conventional evidence of the infectious disease or cancer becomes apparent. This approach is particularly useful for screening blood products for latent pathogens, such as HIV. It is also of immense benefit for the rapid identification of pathogens in acute infections, as this allows the correct regimen of drug treatment to be implemented as soon as possible.

An early example of DNA-based diagnostics was the hybridization test used to detect hemoglobin disorders, which are known as hemoglobinopathies. As discussed above, the globin genes were among the first human genes to be cloned because the cDNA sequences are so abundant. Labeled globin cDNA probes from healthy individuals were hybridized to Southern blots of genomic DNA from both healthy people and those suffering from different hemoglobinopathies. This allowed changes in DNA band patterns that were disease specific to be identified.

Some disease-causing mutations either create or destroy a restriction site, allowing the disease to be diagnosed directly by Southern blot analysis. This occurs in sickle-cell disease, which is caused by a point mutation in the β-globin gene. The mutation destroys the recognition site for the restriction endonuclease *Mst*II, allowing sickle cell individuals (and carriers) to be detected because of the unusually long *Mst*II restriction fragments (Fig. 1.5). In other cases, one or more than one restriction fragments are absent and similar results occur with a number of different restriction endonucleases. This is suggestive of a larger deletion, as occurs in the thallasemias (Fig. 1.5b).

Very few diseases can be diagnosed on the basis of point mutations that change restriction sites, but restriction analysis is unnecessary for mutation detection. If a disease-causing point mutation can be identified, synthetic oligonucleotides can be made corresponding to both the normal and mutant sequences. These **allele-specific oligonucleotides (ASOs)** can be used in two ways. Longer ASOs can be used for **allele-specific hybridization**, a procedure in which the ASOs are labeled and hybridization conditions are adjusted to accept only perfect matches between such oligonucleotides and the target genomic DNA. Alternatively, shorter ASOs can be used as primers in an **allele-specific PCR**. In this case, the last nucleotide of the primer is chosen as the discriminant position because extension will not occur from a primer with a mismatched 3′ end (Fig. 1.6).

The production of therapeutic proteins

The modification of a cloning vector to include regulatory elements that control gene expression allows the cloned gene to be expressed as a **recombinant protein**.

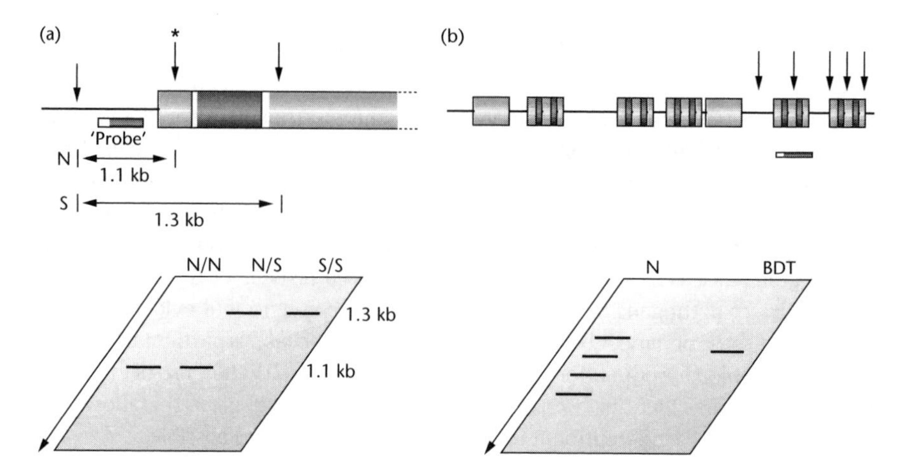

Fig. 1.5 DNA sequences as diagnostic tools. (a) Disease diagnosis by testing for point mutations that alter the number of restriction sites using sickle cell anemia as an example. The top panel shows the human β-globin gene (the gray box represents the coding region and the first intron is shown with darker shading). Vertical arrows represent *Mst*II restriction sites. In normal individuals, there are three sites and the probe will identify a fragment of genomic DNA 1.1 kb in length. The mutation responsible for the disease (*) destroys the central restriction site so that the probe detects a 1.3-kb fragment instead. The lower panel shows a Southern blot from normal (N/N), heterozygous (N/S), and sickle cell disease (S/S) individuals. The arrow shows the direction of electrophoresis. Note the similarity of this technique to the detection of RFLPs (see p. 25). (b) Disease diagnosis by testing for deletions that remove restriction fragments. The top panel shows the β-globin cluster with the genes and pseudogenes identified. The vertical arrows show *Eco*R1 restriction sites in the β-globin and δ-globin genes. The lower panel shows the result of a Southern blot experiment. In normal individuals (N), a β-globin cDNA probe (bar) would reveal several fragments because cross-hybridization to the δ-globin gene would be possible under reduced stringency conditions. In individuals with βδ-thallasemia (BDT) these two genes are deleted, and hybridization to any residual fragments between the outer restriction sites would result in a single hybridizing band. The same result would be expected for other restriction enzymes, e.g. *Hin*dIII. Note the similarity of this technique to loss of heterozygosity mapping in cancer (see p. 118).

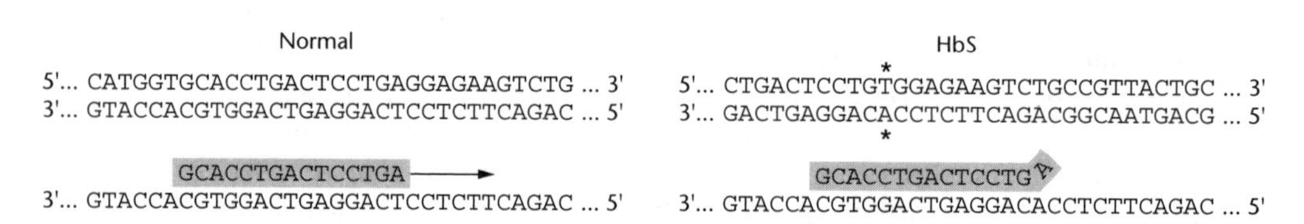

Fig. 1.6 Allele-specific PCR to detect sickle cell anemia. The top panels show the normal and mutant β-globin sequences, with * marking the position of the mutation. The lower panel shows amplification with a PCR primer matching the normal sequence. It will be extended on a normal template (left) but not on a mutant template because the final nucleotide does not anneal (right).

There are many basic applications of this technology including, as discussed above, the use of expression libraries for gene isolation. In medicine, however, the primary application of expression technology is the production of recombinant therapeutic proteins.

Human proteins as drugs

Therapeutic protein synthesis was one of the first commercial applications of recombinant DNA technology and the initial products were simple proteins, like human growth hormone and insulin, for which there was a large demand and an unsatisfactory source. In many cases the authentic product had to be isolated from human cadavers or animals and there was a risk of contamination with pathogens. For example, some children treated with growth hormone extracted from human pituitary glands later developed Creutzfeld–Jakob disease, and many patients treated with human blood products have since developed hepatitis or HIV infections.

The first recombinant proteins were produced in bacteria in the late 1970s and large scale bacterial fermentation continues to be used today. However, while this approach is suitable for simple proteins, bacteria do not carry out many forms of protein post-translational modification, including glycosylation. Alternative systems are thus required for the production of complex glycoproteins. There have been some successes with yeast and insect cells, but the glycan chains added to recombinant proteins are radically different to those produced in mammals. Therefore, many complex recombinant human proteins are produced in large scale cultures of mammalian cells. Because this is very expensive, alternative production systems have been explored and the use of transgenic animals and plants is increasing in popularity. This topic is discussed in more detail in Chapter 6.

Recombinant vaccines

The prevention of infectious diseases by vaccination has a long and successful history beginning in 1796 when Edward Jenner injected a young boy with cowpox, thus conferring protection against a subsequent infection with the deadly smallpox virus. Most of the vaccines in use today are based on similar principles and are known as "Jennerian vaccines." These include live but attenuated bacteria or viruses which cause the body to mount a protective immune response against the target pathogen (e.g. the measles, mumps, rubella, and tuberculosis vaccines) and "killed vaccines," i.e. the pathogen itself is killed so it is no longer infectious but it can still stimulate the immune system.

Unfortunately, vaccines against all common diseases cannot be made using the above methods and other approaches are needed. An alternative strategy is the use of **recombinant subunit vaccines**, where the gene for one specific protein on the pathogen is expressed, and the protein used as the vaccine. The current hepatitis B and influenza vaccines are protein subunits produced in yeast. Since these inert subunits do not multiply inside the vaccinee, they do not generate an effective cellular immune response. To address this, heterologous antigens have been expressed in attenuated bacteria and viruses and used as surrogate live vaccines. For example, vaccinia virus has been used to express a wide range of proteins from different pathogens, including the rabies glycoprotein, leading to the eradication of rabies in some parts of Europe. More recently, genetically transformed plants have been used to produce oral vaccines which can be administered either by eating the plant material directly, or after minimal processing. Vaccines are discussed further in Chapter 3.

The special status of recombinant antibodies

Antibodies bind to target antigens with great specificity and are therefore used in molecular biology for the detection, quantification and purification of proteins. In medicine, antibodies can be used to prevent, detect and cure diseases. For example, antibodies against the surface adhesin of the oral pathogen *Streptococcus mutans* are being developed as a drug to prevent tooth decay, and antigens that recognize specific tumor antigens can be used to diagnose and treat cancer. The traditional way to produce monoclonal (single target specificity) antibodies is to fuse B lymphocytes from immunized mice with immortalized myeloma cells, resulting in the recovery of **hybridoma cell lines** that produce the same antibody indefinitely. The disadvantage of murine antibodies is their immunogenicity in humans. Recombinant DNA technology has been used to address this problem in a number of ways, including the production of humanized antibodies, recombinant antibody derivatives, and antibody fusion proteins. Furthermore, artificial immune diversity can be generated using libraries of antibody variable regions as in phage antibody display. Recombinant antibodies are discussed in Chapter 6.

Gene medicine

Traditionally, DNA sequences have been used to detect diseases while proteins and other "small molecule" drugs have been used to treat or prevent them. This distinction is becoming blurred, however, with the development of novel forms of therapy known collectively as **gene medicine** (see Chapter 8). One form of gene medicine is known as **gene therapy** and involves the introduction of DNA sequences into human cells either *in vitro* or *in vivo* with the purpose of treating and hopefully curing disease. In most cases, gene therapy is directed at diseases caused by mutations in human genes (inherited disorders, cancer) and ideally is meant to alter the genome and provide a permanent cure. In contrast to the use of drugs to alleviate disease symptoms, therapeutic DNA has the capability of correcting the actual cause of the disease by correcting or compensating for the mutation itself. Other forms of gene medicine are more similar to traditional drugs. They include the use of synthetic oligonucleotides, ribozymes, and most recently RNA interference to block the expression of particular mutant genes in the treatment of cancer or infectious diseases. For example, several gene therapy trials are underway which involve various strategies to combat HIV.

A special category of gene medicine is the use of **DNA vaccines**. These are constructs containing the gene corresponding to a pathogen antigen. When expressed in the human body, the antigen is made and induces an immune response providing protection against subsequent infections. DNA vaccines are advantageous because the same strategy can be used to prepare vaccines against many different diseases, and because vaccines against new disease isolates can be developed rapidly. There are also logistic advantages in that DNA is easier to store and transport than proteins.

Disease models

Another major application of recombinant DNA technology is the introduction of predefined mutations into genes by *in vitro* mutagenesis followed by the transfer of such altered genes back into the source organism for functional testing. It is not possible to do this with human genes for ethical reasons, but **disease models** can be created by mimicking human pathogenic mutations in other animals. Such models can be used to investigate the molecular basis of the disease and, importantly, to test novel drugs before clinical trials in humans.

Mammals have been used as human disease models for many years, but until comparatively recently this relied on the identification of spontaneous mutants or the screening of mutagenized populations to identify those with disease-like phenotypes. Recombinant DNA technology in combination with advances in mammalian gene transfer techniques has made it possible to create **exact replicas** of human pathogenic mutations by integrating dominantly malfunctioning transgenes or replacing the endogenous gene with a nonfunctional copy, a technique commonly described as "gene knockout." More recently, it has been possible to model more complex diseases in mice by simultaneously introducing mutations into two or more genes.

The impact of genomics on medicine

The recombinant DNA revolution provided us with tools and techniques to isolate and characterize individual genes, but this approach has two major limitations. First, finding genes one at a time is extremely laborious and expensive work. Second, it encourages a **reductionist** approach to biomedical research, whereas it is well known that genes do not function in isolation. Thousands of genes must work together to coordinate the biologic activities that form a functioning human, or indeed any other organism. The second modern revolution in medicine, the **genomics revolution**, has addressed these drawbacks by encouraging a new **holistic** approach in which genes and their products are characterized in large numbers. Genomics is the study of entire genomes, incorporating mapping, sequencing, annotation (gene finding), and functional analysis. The tools and techniques provided by the genomics revolution are high-throughput equivalents of those from the recombinant DNA era, allowing more data to be gathered and analyzed in a much shorter space of time.

The genomic revolution began in the early 1990s when the Human Genome Project began to gather pace. The initial aims of the project were to map and sequence the entire human genome, leading to the identification of all human genes. The first phase of the project involved the creation of a high-density genetic map that could be used as a framework or scaffold to assemble a physical map of DNA clones. These clones were then sequenced, systematically, and the sequences analyzed for genes. Technical innovations were required in all areas to achieve these aims but the most impressive advances came in the automation of DNA sequencing, which increased the rate of data production over 1000-fold compared to the 1980s. Technology improvements were stimulated by competition from the private sector, and during the progress of the Human Genome Project, the genomes of many bacteria and some eukaryotes were also sequenced. These included many human pathogens and a handful of important model experimental organisms, such as the fruit fly (*Drosophila melanogaster*), the nematode worm (*Caenorhabditis elegans*), and the humble baker's yeast (*Saccharomyces cerevisiae*). We will not consider the methodology of genome mapping and sequencing here since this subject is explored in more detail in Chapter 2.

The output of the first phase of the Human Genome Project was a draft sequence extensively annotated with genes (a **transcript map**). The transcript map is the key to the potential medical benefits of the project because with further refinement it could provide access to all human genes. Therefore, while one of the first benefits of recombinant DNA was access to individual human genes, one of the first benefits of genomics was access to **all of them**. The transcript map is helping to accelerate the rate at which disease genes are discovered because it is now no longer necessary to devise elegant cloning strategies. Positional cloning is obsolete, because once a disease gene has been mapped to a particular genomic region, the transcript map can be inspected for candidate genes and these can be studied for evidence of disease association.

As well as large scale methods for gene isolation, the genomics revolution has also provided large scale methods for functional analysis. Indeed it seems impossible to read about genomics without the phrases "large scale" or "high-throughput" or "massively parallel" being used to describe the experimental methods. The emphasis of genomic technology is on maximizing the amount of data output while minimizing the amount of hands-on input through extensive automation, miniaturization, and parallelization. These techniques are described only very briefly below because they are discussed in more detail in the following chapter. However, compare the list below to the one on page 10:

- Analysis of gene expression. High-throughput expression analysis by large scale cDNA sequencing, sequence sampling techniques and the use of DNA microarrays allows the expression of thousands of genes to be analyzed simultaneously. This can show the global effect of different conditions on gene expression profiles, help to link genes into similar expression (**synexpression**) classes, and home in on differentially expressed genes.

• Analysis of protein expression. High-resolution separation techniques such as two-dimensional gel electrophoresis can be used to fractionate complex protein mixtures, and mass spectrometry can be used to identify individual proteins rapidly and accurately. The expression of thousands of proteins can be analyzed and compared across samples.
• Analysis of protein interactions. New high-throughput technologies such as phage display, the yeast two-hybrid system and mass spectrometric analysis of protein complexes allow interacting proteins to be cataloged on a large scale. Protein interaction maps of whole cells can be produced.
• Altering gene expression or activity. Large scale mutagenesis can be used to generate populations with either random or targeted mutations in every single gene. Similarly, RNA interference can be applied on a large scale to inactivate all the genes in the genome systematically. Mutation techniques can be applied only to model organisms but RNA interference is used in human cells.
• Analysis of protein structure. Large scale "structural genomics" programs have been initiated to solve many protein structures. It is hoped that representatives of all protein families will be structurally solved to increase the rate at which functions are assigned to genes.

Advances in **bioinformatics** (the use of computers to process biologic data) have gone hand in hand with advances in genomics because only computers have the power to analyze the large datasets produced by genomic-scale experiments. One of the most important contributions of bioinformatics is **sequence analysis**, which allows sequences of genes and whole genomes to be compared. There is extensive structural and functional conservation among genes and even whole molecular pathways between humans and model organisms such as the fruit fly, the nematode worm, and the baker's yeast. Up to 20% of human disease genes have counterparts in yeast and up to 60% have counterparts in the worm and fly, allowing these organisms to be used for functional analysis and the screening of candidate drugs. Similarly, comparisons between bacterial sequences, especially those of harmless species and related pathogens, are helping to reveal virulence factors and pathogenesis-related proteins that could be used as new drug targets or candidates for new vaccines. Another important role of bioinformatics is the presentation of data in easily accessible and user-friendly databases, allowing the efficient dissemination of information. As we shall see later in the book, some databases are already having a real impact on our understanding of disease at the molecular level, and this will have a knock-on effect on the development of novel therapies. One example is the Cancer Genome Anatomy Project, which aims to assemble gene expression and functional data from all forms of cancer.

The new molecular medicine

The potential availability of all human disease genes, as well as genes in human pathogens that are responsible for infectious diseases, is likely to have a major impact on drug development. At the current time, most available drugs interact with a small repertoire of 500 or so target proteins in the body. There are approximately 30,000 genes in the human genome and many of these will represent novel drug targets. Therefore, the functional analysis of these genes and the structural analysis of their products could lead to an explosion in the number of drugs being developed in the next few decades. Furthermore, the growing recognition of the importance of conserved molecular pathways and the tendency of proteins to function in large complexes will allow key regulatory molecules to be selected as drug targets. Pharmaceutical companies have not been slow to embrace the potential of genomics, and we discuss the process of drug development in Chapter 7.

Another aspect of genomics that is likely to have a large impact on medicine is the analysis of human **variation**. Earlier in this chapter, we discussed the use of DNA sequences as diagnostic tools to identify particular sequence variants associated with disease. More recently, techniques based on the same principles have been streamlined and miniaturized for the high-throughput analysis of **single nucleotide polymorphisms (SNPs)**. Unlike **disease-causing** point mutations, SNPs are common variants that are widespread in the population. While they do not **cause** overt diseases, some are thought to **contribute** in a small and additive manner to disease susceptibility, and to other complex characteristics such as individual responses to drugs. Spin-offs from the Human Genome Project aim to catalog all the SNPs in the genome (there are thought to be 10 million in total, with any two individuals varying at about 3 million positions) as well as blocks of SNPs, known as **haplotypes**, that are tightly linked and tend to be inherited as a group. For the first time, it may be possible to pinpoint the genetic variants that predispose us to common diseases, such as asthma and diabetes (see Chapter 4). It may also be possible to identify genetic variants that influence our responses to drugs, raising the possibility of personalized medicines targeted to the genetic composition of individual patients (see Chapter 7). We must be careful, however, to guard against the misuse of genetic information arising from the Human Genome Project and its subsidiaries. A large segment of the budget for this project has been set aside to address the social, legal and ethical issues involved, in order to protect the privacy of those contributing their DNA to the project and to prevent data from human genomic analysis being used to discriminate against individuals or ethnic groups.

Outline of this book

The aim of this book is to provide a broad and comprehensive account of how recombinant DNA technology and genomics are used in medicine. The next chapter explains the principles of genomics in enough detail for the reader to understand the material presented in later chapters. Chapters 3–5 discuss the role of recombinant DNA and genomic analysis in the diagnosis, treatment and prevention of infectious diseases, inherited diseases, and cancer. The subsequent three chapters cover emerging types of therapy and modern approaches to drug development. A "roadmap" of the book is shown in Fig. 1.7.

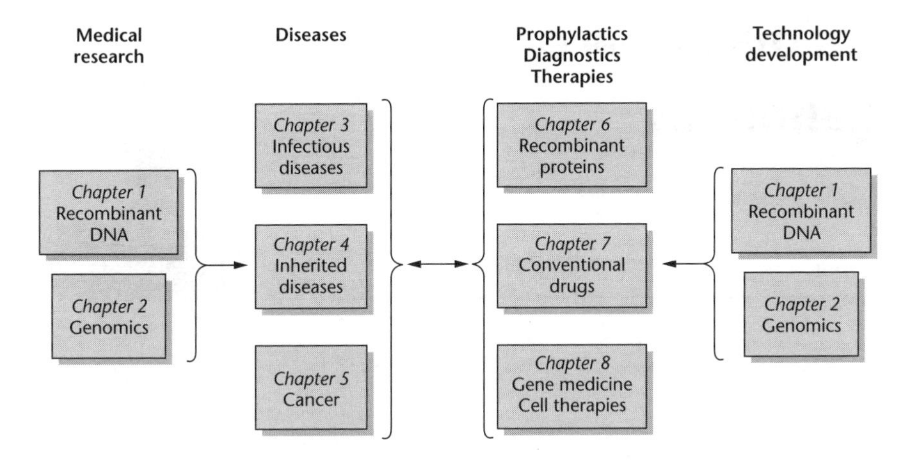

Fig. 1.7 A "roadmap" of the layout of this book.

Further reading

POGM: Chapter 1 provides an overview of recombinant DNA technology and describes the birth of the biotechnology industry. Chapter 2 introduces basic techniques while Chapters 3–6 discuss cloning vectors and strategies in more detail. Chapter 14 has sections on the applications of recombinant DNA technology in medicine.

POGA: Chapter 1 introduces genomics and some of its applications. Chapter 12 has sections on the applications of genomics in medicine.

Williams SJ, Hayward NK (2001) The impact of the Human Genome Project on medical genetics. *Trends Mol Med* **7**, 229–231.

Yaspo M-L (2001) Taking a functional genomics approach in molecular medicine. *Trends Mol Med* **7**, 494–502.

Two useful articles, one a summary and one an in-depth review, discussing the impact of genomics on molecular medicine.

Wren BW (2000) Microbial sequencing: insights into virulence, host adaptation and evolution. *Nature Rev Genet* **1**, 30–38.

A thorough article showing how microbial genomics is providing new leads in the fight against infectious disease.

C H A P T E R T W O

An overview of genomics

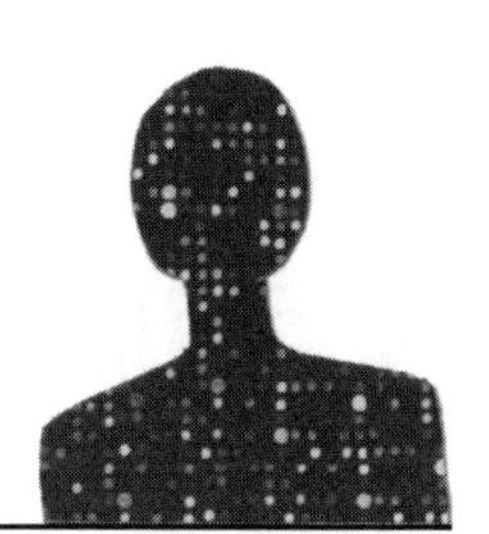

Introduction

In the previous chapter, we charted the history of molecular medicine from its origins in the aftermath of the recombinant DNA revolution to the present day, and briefly discussed some of the expected scientific and medical benefits of genomics. The position we are in now is one of enormous promise. At our fingertips, we have the complete sequence of the human genome and potential access to every single gene. This offers an unprecedented opportunity to study human biology, in health and disease, in a truly global and systematic way. Similar resources are available for a large number of other organisms of medical relevance, including some of our most important pathogens (Table 2.1). The focus of medical research is now turning to the systematic functional evaluation of genes and the elucidation of pathways and networks. A complete understanding of how genes function and interact to co-ordinate the biologic activities that make a healthy human provides enormous

Table 2.1 Some pathogen genomes (bacterial and protozoan) that have been sequenced.

Pathogen	Disease	Genome size (Mb)
Bacillus anthracis	Anthrax	4.5
Bordetella pertussis	Whooping cough	3.88
Borrelia burgdorferi	Lyme disease	0.95
Helicobacter pylori	Peptic ulcers	1.67
Leishmania major	Leishmaniasis	33.6
Mycobacterium leprae	Leprosy	2.8
Mycobacterium tuberculosis	Tuberculosis	4.4
Plasmodium falciparum	Malaria	23
Rickettsia prowazekii	Typhus	1.1
Salmonella typhi	Typhoid fever	4.5
Treponema pallidum	Syphilis	1.1
Trypanosoma brucei	Sleeping sickness	54
Vibrio cholerae	Cholera	2.5
Yersinia pestis	Plague	4.38

scope for the development of novel therapies. In this chapter, we review the scientific achievements that have led us to our current position and consider some of the emerging genomic technologies that may provide medical breakthroughs in the future.

A review of progress: the Human Genome Project

Genomics (Box 2.1) became a significant and independent field of research in 1990 when the **Human Genome Project (HGP)** was officially launched. The stated aim of the project was to sequence the entire 3000-Mb human nuclear genome within 15 years. At the outset, however, it was acknowledged that a great deal of preliminary work was required before actual sequencing could begin, and that five model organism genomes should be sequenced in addition to the human genome to act as pilot projects for the validation of new technologies (Box 2.2). One of the first tasks was to construct a high-resolution genetic map of the human genome to act as a scaffold for the assembly of a physical map of DNA clones. Once the genetic and physical mapping phases were completed, then sequencing could begin. Technological advances were required in mapping, cloning, sequencing,

Box 2.1 What is genomics?

The term **genome** was introduced in 1920 by the German botanist Hans Winkler to describe the collection of genes contained within a complete (haploid) set of chromosomes. Nowadays, the term has expanded to include all the DNA in a haploid set of chromosomes, not just the genes, because in higher eukaryotes genes are in the minority. For example, only 2–3% of the human genome is represented by genes. Although the concept of the genome is longstanding, the term **genomics** was not used for the first time until 1986. The mouse geneticist Thomas Roderick introduced this word to describe the mapping, sequencing and characterization of genomes. More recently, the essence of genomics has become associated with any form of large scale, high-throughput biologic analysis and has spawned a whole lexicon of derivative terms. **Functional genomics** encompasses any systematic approach to the analysis of gene function, and many of the technologies of functional genomics are discussed in this chapter. **Transcriptomics** is the large scale analysis of mRNA expression. **Proteomics** is the large scale analysis of proteins, and can itself be divided into the study of expression profiles, interactions, and protein structure. Proteomics is a very significant component of the new molecular medicine because most drug targets are proteins.

Box 2.2 Model organism genomes as initial targets of the Human Genome Project

Escherichia coli (bacterium)
Saccharomyces cerevisiae (yeast)
Caenorhabditis elegans (nematode)
Drosophila melanogaster (fruit fly)
Mus musculis (mouse)

Box 2.3 The ethical, legal and social issues (ELSI) of the Human Genome Project

Before the Human Genome Project was inaugurated, it was recognized that both the way in which the project was carried out and the data it produced would raise new and complex ethical issues. Particular areas of concern included matters relating to the collection of samples, the privacy of donors, and the availability and subsequent use of genetic information arising from the project. Therefore, both of the US organizations sponsoring the HGP – the US Department of Energy (DOE) and the National Institutes of Health (NIH) – devoted a significant proportion of their annual HGP budgets (3% and 5% respectively) to fund a series of programs whose aim was to study the ethical, legal and social issues (ELSI) of the project. The function of the ELSI programs was, and is, to promote education and guide policy decisions by consultation with a wide range of interested parties. A unique aspect of the HGP ELSI programs is that they are integral to the project itself rather than retrospective, and therefore help to foresee the implications of new technology developments and address any important issues before problems arise.

The initial aims of the ELSI programs were stated as follows:

- To anticipate and address the implications for individuals and society of mapping and sequencing the human genome
- To examine the ethical, legal and social consequences of mapping and sequencing the human genome
- To stimulate public discussion of the issues, and
- To develop policy options that would assure that the information is used for the benefit of individuals and society.

In the 10 years since the ELSI programs were initiated, a large body of work has been produced to educate policymakers and the public. This has helped in the development of policies relating to the conduct of genetic research and the commercial exploitation of genetic information and its associated technologies. Some of the more important challenges relate to the spin-off projects that focus on human genetic variation, i.e. the SNP mapping project and the haplotype mapping project. In these cases the privacy of individuals and communities contributing DNA samples must be protected, but it is also necessary to obtain informed consent and to provide continuous liaison through advisory groups. A major concern is that information on genetic variation could be used to discriminate against individuals or populations in terms of employment, insurance, or legislation. ELSI programs have been established to anticipate how these data may affect concepts of race and ethnicity and to foresee the impact of technologic advances and data availability on the entire concept of humanity. The educational resources not only help to keep the public and policymakers informed, but also help scientists to present their results carefully to avoid misinterpretation.

The aims of ELSI are updated every few years and the most recent are presented below:

- To examine issues surrounding the completion of the human DNA sequence and the study of human genetic variation
- To examine issues raised by the integration of genetic technologies and information into health care and public health activities
- To examine issues raised by the integration of knowledge about genomics and gene–environment interactions in nonclinical settings
- To explore how new genetic knowledge may interact with a variety of philosophical, theological and ethical perspectives
- To explore how racial, ethnic and socioeconomic factors affect the use, understanding and interpretation of genetic information, the use of genetic services, and the development of policy.

and bioinformatics, in order to achieve the goals of the HGP within the allotted time frame. A large part of the initial budget was also set aside to address the **ethical, legal and social issues (ELSI)** that arose from the project, such as preventing any data arising from the project being used to discriminate against individuals or populations (Box 2.3).

To place the ambitious technical objectives of the HGP in context, consider that in the mid-1980s when the project was first conceived, it was possible to sequence about 1000 nucleotides of DNA per day. At that rate, armies of scientists doing nothing but sequencing would have been required to complete the whole genome. Sydney Brenner, one of the proponents of large scale biology, joked that sequencing should be done by prisoners! It was envisaged that entirely new sequencing methods would be needed in order to increase data output to the required levels. However, although several new methods emerged during the HGP, the goal of increased output was met in the most part by the automation and multiplexing of existing technology. Using ultrarapid capillary sequencers that process 96 samples at once, it is now possible to produce upwards of half a million nucleotides of sequence per day with one machine. Further multiplexing, and the use of multiple machines, can increase this output even more.

Breakthroughs in genetic mapping

Genetic maps are based on **recombination frequencies**, and in model organisms they are constructed by carrying out large scale crosses between different mutant strains. The principle of a genetic map is that the further apart two loci are on a chromosome, the more likely that a crossover will occur between them during meiosis. Recombination events resulting from crossovers can be scored in genetically amenable organisms such as *Drosophila* and yeast by looking for new combinations of the mutant phenotypes in the offspring of the cross. This approach cannot be used in human populations because it would involve setting up large scale matings between people with different inherited diseases. Instead, human genetic maps rely on the analysis of **DNA sequence polymorphisms** in existing family pedigrees (Box 2.4).

Prior to the HGP, low-resolution genetic maps had been constructed using **restriction fragment length polymorphisms (RFLPs)**. These are naturally occurring variations that create or destroy sites for restriction enzymes and therefore generate different sized bands on Southern blots (Fig. 2.1). The problem with RFLPs was that they were too few and too widely spaced to be of much use for constructing a framework for physical mapping – the first RFLP map had just over 400 markers and a resolution of 10 cM, equivalent to one marker for every 10 Mb of DNA. The necessary breakthrough came with the discovery of new polymorphic markers, known as **microsatellites**, which were abundant and widely dispersed in the genome (Fig. 2.2). By 1992, a genetic map based on microsatellites had been constructed with a resolution of 1 cM (equivalent to one marker for every 1 Mb of DNA) which was a suitable template for physical mapping. However, efforts in genetic mapping did not stop there. By 1996 a further map incorporating additional microsatellite markers was published, with a resolution of 0.5 cM. The most recent map, released in 2002 by the deCODE consortium in Iceland, has a resolution of 0.2 cM and incorporates over 5000 markers. The SNP and haplotype projects are also examples of high-resolution genetic maps (Box 2.4).

Box 2.4 Variation in the human genome

The DNA used for the HGP came from 12 anonymous volunteers. Since the genome sequences of any two unrelated humans are only 99.9% identical, there is no "correct" sequence. However, it is the 0.1% difference – amounting to 3 million base pairs of DNA – which is the most interesting, as this makes each of us unique. **Gene mutations** that cause inherited diseases are very rare in the population as a whole and therefore account for only a tiny proportion of this variation. The vast majority occurs in the form of **sequence polymorphisms**, where several different variants (**alleles**) may be quite common. These variations are used as markers to create genetic maps because hybridization or PCR assays (see Chapter 1) can be used to detect and identify the alleles and therefore establish whether recombination has occurred in a family pedigree.

Types of variation

About 95% of polymorphic sequence variation is represented by **single nucleotide polymorphisms (SNPs)**, i.e. single nucleotide positions that may be occupied by one base in some people but an alternative base in others. Where these polymorphisms occur in and around genes, they may occasionally have overt phenotypic effects (e.g. polymorphisms affecting hair color). In most cases, however, the effects of SNPs are far more subtle, e.g. they may influence in a small but additive manner our disease susceptibility or response to certain drugs (see p. 108). The vast majority of SNPs occur outside genes and probably have no effect. However, they are still useful as genetic markers. Some SNPs either create or destroy restriction enzyme sites, so altering the pattern of bands seen on a Southern blot. These **restriction fragment length polymorphisms (RFLPs)** were used to produce the first comprehensive genetic map of the human genome.

The remaining 5% of sequence polymorphism occurs mostly in the form of **simple sequence repeat polymophisms (SSRPs)** otherwise known as **microsatellites**. These are short sequences repeated a variable number of times. The most common form of microsatellite is $CA(_n)$, where n represents the number of repeats (typically 5–50). Unlike SNPs, microsatellites have multiple alleles (i.e. there may be common variants with 12 repeats, 22 repeats, 31 repeats, etc.) whereas SNPs usually occur as one of two alternative forms. Microsatellites rarely occur within genes, and often have pathogenic effects when they do (e.g. Huntington's disease), but they are widely distributed and can be used to produce a much higher resolution map than RFLPs. The physical mapping stage of the Human Genome Project used as a scaffold a genetic map based on microsatellite markers.

Studying variation

Human variation has been used in forensic analysis for many years but interest in **genome-wide variation** began to grow only as the HGP gathered pace. A global effort to study human sequence diversity, **the Human Genome Diversity Project (HGDP)**, was initiated as a spin-off project from the HGP in 1991. However, it received little funding because the primary aim of the project was to find markers corresponding to different ethnic groups for the study of population history and human origins. There has been much more support for SNP mapping projects, both public and private, since these provide concrete benefits to medical research. The ability to identify associations between SNPs and disease susceptibility should greatly accelerate the rate at which disease genes are discovered, and associations between SNPs and drug responses underlie the new medical field of **pharmacogenomics**, where drugs can be tailored to individuals based on their genotype (see Chapter 4). **The International SNP Consortium Ltd** started a systematic SNP mapping project in 1999 and had produced a map containing nearly one and a half million SNPs by 2001. More recently, it has been shown that groups of SNPs tend to be inherited together as **haplotype blocks** with little recombination within them. The estimated 10 million SNPs could therefore be represented by as few as 200,000 haplotypes which would make the process of establishing disease associations much easier. An **International HapMap Project**, aiming to map haplotypes throughout the genome, was inaugurated in October 2001.

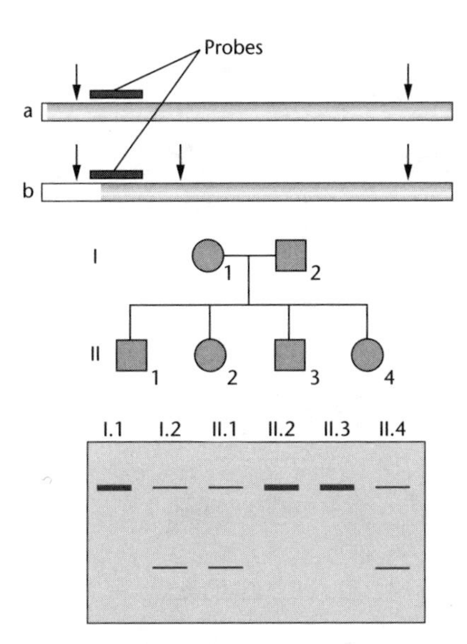

Fig. 2.1 Restriction fragment length polymorphisms (RFLPs) are sequence variants that create or destroy a restriction site therefore altering the length of the restriction fragment detected by a given probe. The top panel shows two alternative alleles, in which the restriction fragment detected by a specific probe differs in length due to the presence or absence of the middle one of three restriction sites (represented by vertical arrows). Alleles a and b therefore produce hybridizing bands of different sizes in Southern blots (lower panel). This allows the alleles to be traced through a family pedigree. For example, child II.2 has inherited two copies of allele a, one from each parent, while child II.4 has inherited one copy of allele a and one of allele b. Note the similarity of this method to the detection of disease alleles such as the sickle cell disease variant of β-globin (Fig. 1.5). Essentially, the only difference is that RFLPs are more common in the population than disease-related mutations because they do not have overt and striking effects on the human phenotype.

Breakthroughs in physical mapping

Unlike genetic maps, **physical maps** are based on real units of DNA and therefore provide a suitable basis for sequencing. The physical mapping phase of the HGP involved the creation of genomic DNA libraries (see Chapter 1) and the identification and assembly of overlapping clones to form **contigs** (unbroken series of clones representing contiguous segments of the genome). When the HGP was initiated, the highest-capacity vectors available for cloning were **cosmids**, with a maximum insert size of 40 kb. Because hundreds of thousands of cosmid clones would have to be screened to assemble a physical map, there was an immediate need for **large-insert cloning vectors** which would reduce the amount of work involved. New approaches were also required to find overlaps and assemble clone contigs on the genomic scaffold.

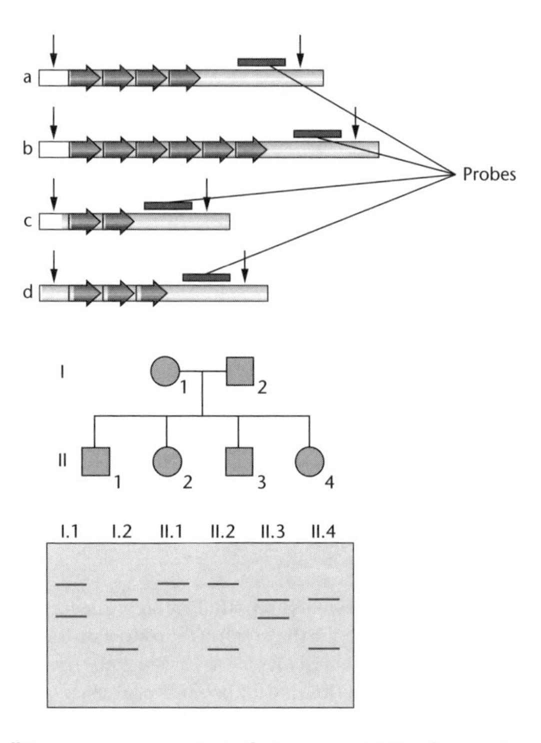

Fig. 2.2 Microsatellites are sequence variants that cause restriction fragments or PCR products to differ in length due to the number of copies of a short tandem repeat sequence, 1–12 nt in length. The top panel shows four alternative alleles, in which the restriction fragment detected by a specific probe differs in length due to a variable number of tandem repeats. All four alleles produce bands of different sizes on Southern blots (lower panel) or different sized PCR products (not shown). Unlike RFLPs, multiple allelism is common for microsatellites so the precise inheritance pattern can be tracked. For example, the mother and father in the pedigree have alleles b/d and a/c respectively (the smaller DNA fragments move further during electrophoresis). The first child, II.1, has inherited allele b from his mother and allele a from his father.

In the case of cloning vector technology, the necessary breakthrough came with the development of **artificial chromosome vectors** that could accept very large inserts (Fig. 2.3). The first such vectors were **yeast artificial chromosomes (YACs)**, which could carry inserts of over 1 Mb reducing the number of clones required to cover the genome to just over 10,000. One problem with YACs, however, was their tendency to incorporate **chimeric inserts** (i.e. inserts comprising segments of DNA from two or more nonadjacent locations in the genome). Therefore, higher-fidelity vectors were required to generate the final physical maps used for sequencing. **BACs (bacterial artificial chromosomes)** and **PACs (P1 artificial chromosomes)** were chosen because of their stability and relatively large insert size (200–300 kb).

Various strategies have been devised to assemble physical clones into contigs, all of which involve the detection of overlaps between adjacent clones. These include:

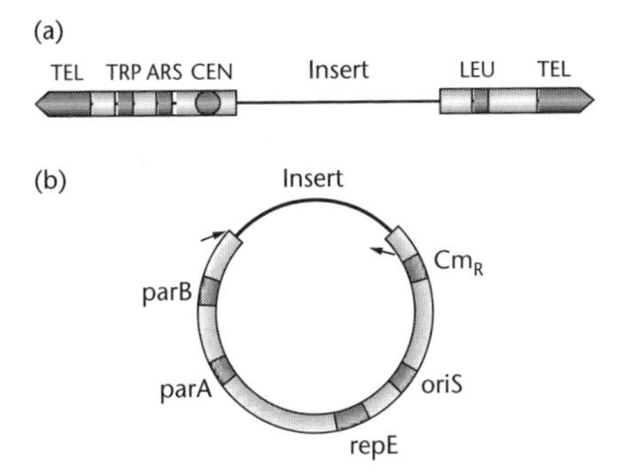

Fig. 2.3 Two artificial chromosome vectors that were invaluable in the human genome project. (a) Yeast artificial chromosome, maximum insert size up to 2 Mb. TEL, telomere; TRP, tryptophan synthesis selectable marker; ARS, yeast origin of replication (autonomous replication sequence); CEN, centromere; LEU, leucine synthesis selectable marker. (b) Bacterial artifical chromosome, maximum insert size up to 200 kb. Cm_R, antibiotic resistance marker; oriS/repE, sequences required for replication; parA/parB, sequences required for copy number regulation. Arrows indicate promoters for T3 and T7 RNA polymerases, which are used to prepare labeled probes corresponding to the end-sequences of the insert.

- **Chromosome walking**. This technique has been widely used for positional cloning (see p. 9) and involves the stepwise use of clones as hybridization probes to identify overlapping ones (see Fig. 1.3). Alternatively, the end-sequences of each clone can be used to design primer pairs and overlapping clones can be detected by PCR.
- **Restriction enzyme fingerprinting**. This technique involves the digestion of clones with panels of restriction enzymes. Two clones that overlap will share a significant number of identical restriction fragments. The patterns are complex and must be interpreted by computers (Fig. 2.4).
- **Repetitive DNA fingerprinting**. As an extension of the above, Southern blots of the restriction fragments can be probed for genome-wide repeat sequences such as *Alu*. There are over a million copies of the *Alu* element dispersed in the genome (one every 4 kb), so a typical 100-kb BAC clone will contain 20–30 repeats. Overlapping clones will share a significant proportion of hybridizing bands. PCR-based fingerprinting tests based on repetitive DNA can also be used.
- **STS mapping**. A **STS (sequence tagged site)** is a unique sequence in the genome, 100–200 bp long, which can be detected easily by PCR. If two clones share the same STS, then by definition they overlap and can be united in a contig.

STS mapping was the most valuable strategy for contig assembly in the HGP because a **physical reference map** containing 15,000 STS markers with an average spacing of 200 kb was published in 1995 (Box 2.5). Therefore, clones containing particular STS markers could be anchored to the reference map to show their precise chromosomal location, not just their relationship to other clones. Importantly, some of the STSs contained polymorphic microsatellite sequences,

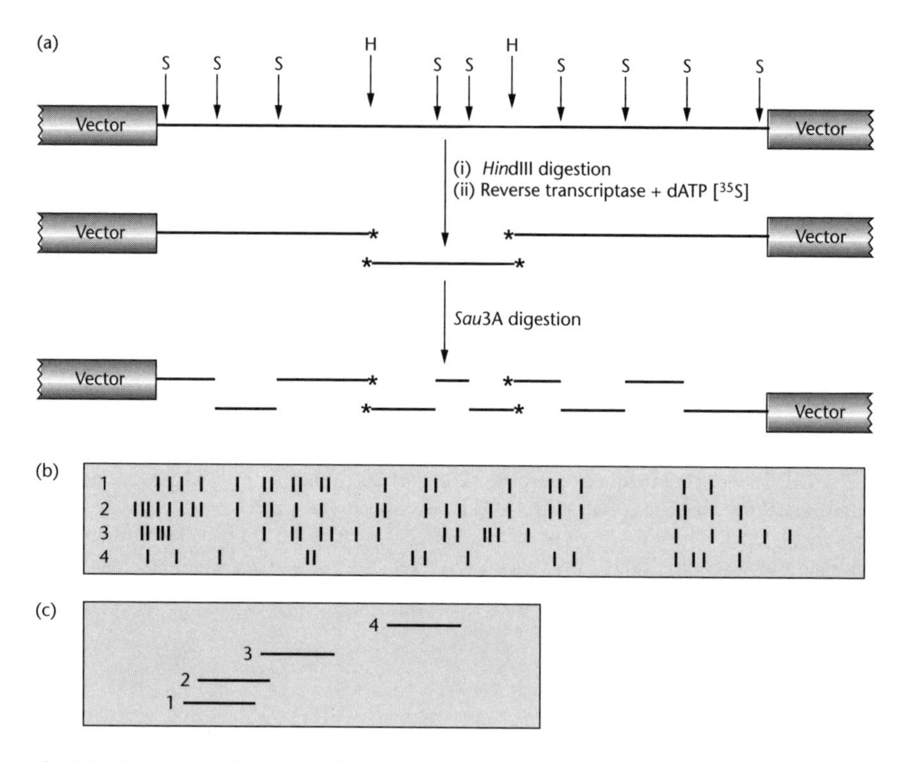

Fig. 2.4 The principle of restriction-fragment fingerprinting. (a) The generation of labeled restriction fragments (see text for details). (b) Pattern generated from four different clones. Note the considerable band sharing between clones 1, 2, and 3, indicating that they are contiguous, whereas clone 4 is not contiguous and has few bands in common with the other three. (c) The contig map produced from data shown in (b).

allowing them to double as genetic markers and integrate with the genetic map. Others were **expressed sequence tags (ESTs)**, derived from cDNA clones, and therefore identified the positions of genes. The importance of ESTs in gene mapping is considered below.

Sequencing strategies

All the cellular genome projects have been based on the fundamental technology of **chain terminator sequencing**, which is explained in Fig. 2.5. Even with the most sophisticated apparatus, however, it is difficult to produce more than 600–700 nucleotides of good sequence per reaction. Therefore, in order to sequence the large insert of a BAC or PAC vector (up to 200 kb), it must be broken down into much shorter segments that are sequenced individually. This is generally achieved by random shearing of the insert into fragments 1–2 kb in length. All the sequence data are then entered into a computer, which can search for overlaps and re-assemble the complete sequence of the original insert using a dedicated assembly algorithm such as PHRAP. This approach is known as **shotgun sequencing**.

Box 2.5 An STS reference map of the human genome

Sequence tagged sites are DNA sequences, 100–200 bp long that are unique in the genome and easily detected by PCR. A physical reference map of the human genome was published in 1995 comprising 15,000 STS markers with an average spacing of 200 kb. This was used as a scaffold to assemble BAC and PAC clone contigs, and as a means to identify overlaps between adjacent clones. But where did the STS markers come from in the first place and how was the map created?

STS markers arose from three sources:

- Some microsatellite markers were borrowed from the genetic map. Microsatellites can double as STS markers so long as they contain some unique DNA in addition to the repeat sequence.
- The random sequencing of clones from cDNA libraries produced partial cDNA sequences known as **expressed sequence tags (ESTs)**. These can be used as STS markers so long as they come from unique genes (as opposed to members of gene families).
- The remainder of the STS markers were derived from unique sequences in random genomic clones.

The tricky issue of mapping the STS markers relative to each other, to create the reference map, was addressed by typing a panel of **radiation hybrids**. This is a classic physical mapping technique in which human cells are lethally irradiated, and individual chromosome fragments are rescued by fusion of the human cells to rodent cells. Panels of cells containing different human chromosome fragments can be tested by PCR for the presence of STS markers. As in genetic mapping, the closer two markers are to each other, the less likely they are to be separated (in this case by chromosome fragmentation rather than crossing over). Therefore, the analysis of many hybrid cells to see which markers were present together on the same chromosome fragment allowed the order of markers to be established (Fig. B2.5a). This was confirmed by testing for the presence of two or more adjacent STS markers in YAC libraries (Fig. B2.5b).

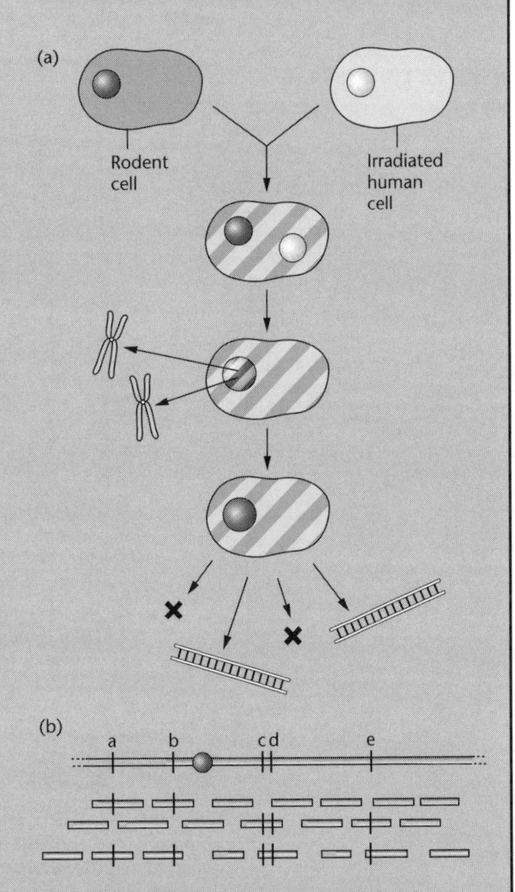

Fig. B2.5 (a) Radiation hybrid mapping. Rodent cells and lethally irradiated human cells can be fused together to generate heterokaryons (cells with two nuclei). These combine to form a hybrid nucleus from which the damaged human chromosomes are eliminated. The result is essentially a rodent cell containing one or more human chromosome fragments. Panels of such hybrids can be created spanning the entire human genome. The systematic testing of such panels for STS markers provides a reference physical map. **(b)** This can be refined and confirmed by testing for the presence of such markers in YAC inserts.

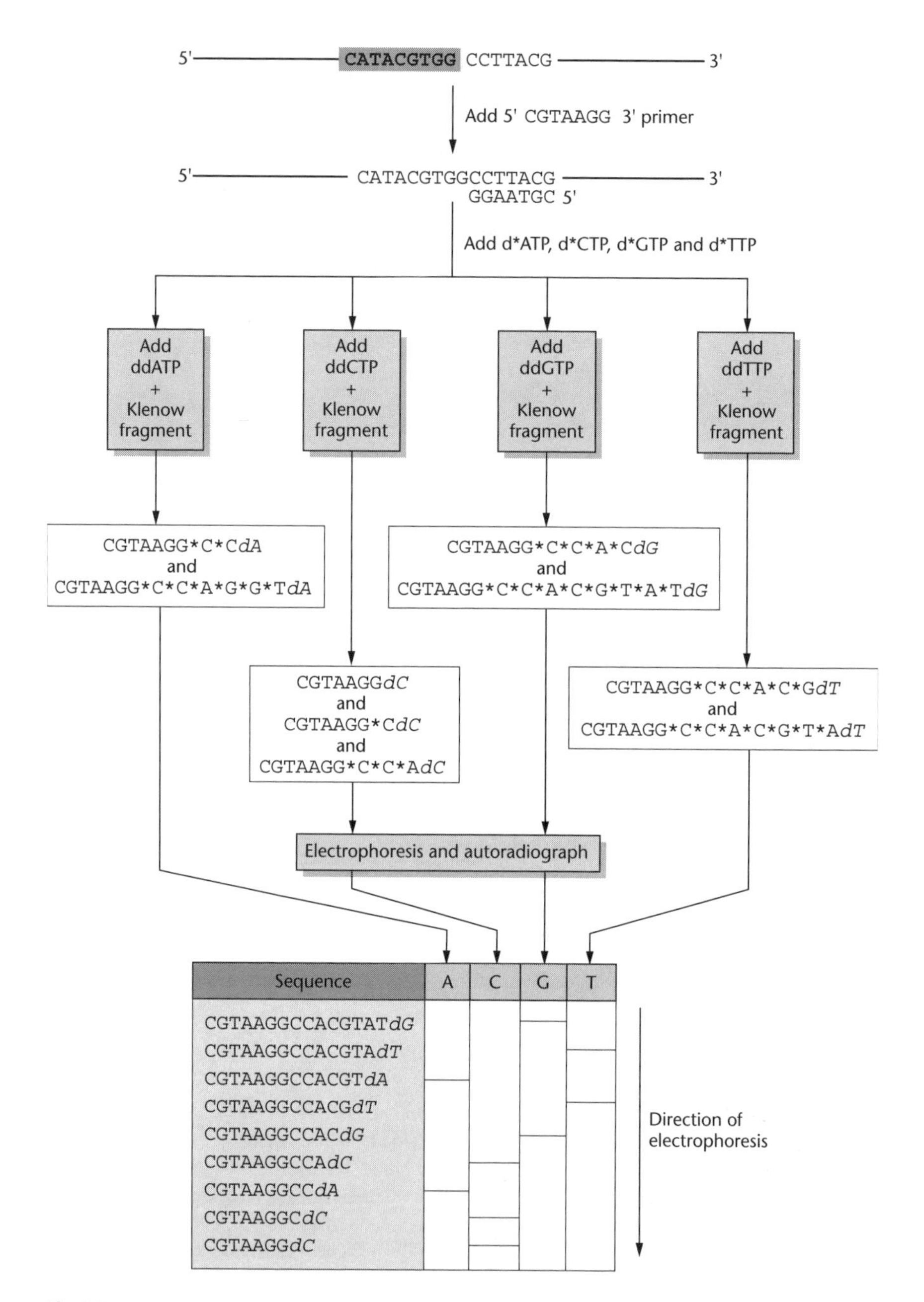

Fig. 2.5 DNA sequencing with dideoxynucleoside triphosphates as chain-terminators. In this figure asterisks indicate the presence of ^{32}P and the prefix "d" indicates the presence of dideoxynucleotide. At the top of the figure the DNA to be sequenced is enclosed within a box. Note also that unless the primer is also labeled with a radioisotope the smallest band with the sequence CGTAAGG*dC* will not be detected by autoradiography as no labeled bases were incorporated.

The HGP employed a **hierarchical shotgun** strategy, in which shotgun sequencing was applied to the inserts of individual BAC clones one at a time. Because each BAC had already been physically mapped at this stage, the position of the sequence on the physical reference map could be determined very easily. In 1999, a privately funded effort to sequence the human genome was launched by the US biotechnology company Celera Genomics, using an alternative **whole genome shotgun** strategy. In this approach, shotgun sequencing is carried out on whole genomic DNA. There is no investment in mapping. Instead powerful computers are used to assemble the entire genome from the short 600–700 nucleotide sequence reads. The project's coordinator, Craig Venter, had used the whole genome shotgun technique to complete the first cellular genome in 1995 and had validated its use on a complex eukaryotic genome by participating in the joint private public project to sequence the euchromatic portion of the *Drosophila melanogaster* genome (Table 2.2). The clone-by-clone and whole genome shotgun methods are compared in Fig. 2.6. Overall, the clone-by-clone approach is slower due to the need for initial mapping and clone assembly stages, but is much easier to finish off because the hierarchical sequence assembly is less demanding in terms of computer resources. Conversely, while the whole genome shotgun method generates data rapidly, the assembly stage is much more of a challenge, particularly due to the abundance or repetitive DNA in the human genome (Fig. 2.7). Indeed it has been suggested that Celera relied on both the maps and the sequence data generated by the HGP (both freely available over the Internet) in order to complete its own draft sequence. The public HGP and Celera jointly announced the completion of the draft sequence in 2000 (publishing reports on the achievement in special issues of the journals *Nature* and *Science* 8 months later) and the sequences were finished in 2003 (Box 2.6).

Genome annotation

The first **postsequencing** task in any genome project is **genome annotation**, i.e. the derivation of useful biologic information from the sequence. Essentially this means finding genes and their regulatory elements that represent the functional components of the genome and have the greatest medical relevance.

There was a strong focus on genes from the very beginning of the HGP, involving the high-throughput sequencing of cDNA clones to generate large collections of ESTs. As discussed above, ESTs are 100–200 bp fragments of cDNA obtained by single-pass sequencing of clones randomly selected from cDNA libraries (rather than the 8–10 reads required for finished sequence). Therefore, although short and inaccurate, ESTs provide a rapid and inexpensive route to the identification of gene sequences as well as being useful physical markers in their own right. About 100,000 ESTs have been mapped onto the genome by typing radiation hybrids and YACs (Box 2.5). Of course, not all these sequences represent individual genes, and there have been attempts to identify nonredundant gene sets by combining overlapping ESTs (e.g. the UniGene Project; http://www.ncbi.nlm.nih.gov/UniGene/). The first comprehensive gene maps were produced in 1996 and provided evidence for 20,000–30,000 genes. At the time, this was considered to be only a small part of the human gene catalog.

Table 2.2 Landmark genome projects.

Year	Organism	Genome size	Comments
1977	Bacteriophage ϕX174	5.38 kb	First genome sequenced. Validation of new sequencing method (chain terminator sequencing) which would come to dominate future genome projects
1995	*Haemophilus influenzae*	1.8 Mb	First cellular genome, first bacterial genome and first human pathogen to be sequenced. Achieved in less than 3 months using the whole genome shotgun method
	Mycoplasma genitalium	0.58 Mb	Smallest known cellular genome
1996	*Saccharomyces cerevisiae*	12 Mb	First eukaryotic genome, important model organism. Inspiring example of an international collaborative effort
	Methanococcus jannaschii	1.66 Mb	First achaean genome
1997	*Escherichia coli*	4.7 Mb	Most important bacterial model species. Sequenced independently by two competing groups
1998	*Caenorhabditis elegans*	97 Mb	First genome of a multicellular organism, and first animal genome, to be sequenced
2000	*Drosophila melanogaster*	165 Mb	Important model organism for human biology, sequenced cooperatively by publicly and privately funded organizations including Celera
	Arabidopsis thaliana	125 Mb	First plant genome
2001	*Homo sapiens*	3000 Mb	The human genome, sequenced independently by the publicly funded HGP and Celera
2002	*Fugu rubripes*	400 Mb	Pufferfish genome, the smallest known vertebrate genome with minimal repetitive DNA, should provide help in the identification of human genes
2003	*Mus musculis*	2800 Mb	The mouse, a model mammal extensively used to study human disease (see Chapter 8). Closest organism to man to be sequenced
	Plasmodium falciparum, *Anopheles gambiae*		The malaria parasite (*P. falciparum*) was the first eukaryotic parasite to be sequenced, an achievement made more significant by the simultaneous publication of the sequence of its vector, the mosquito *A. gambiae*

When the genome sequence became available in 2001, it was expected to reveal a substantial number of new genes. To everyone's surprise, the total gene number was much lower than expected. Current estimates suggest we possess fewer than 30,000 genes, which is only 50% more than the nematode *Caenorhabditis elegans*. The precise number cannot be established with confidence because some genes may be difficult to identify or delineate with accuracy. A gene is predicted if:

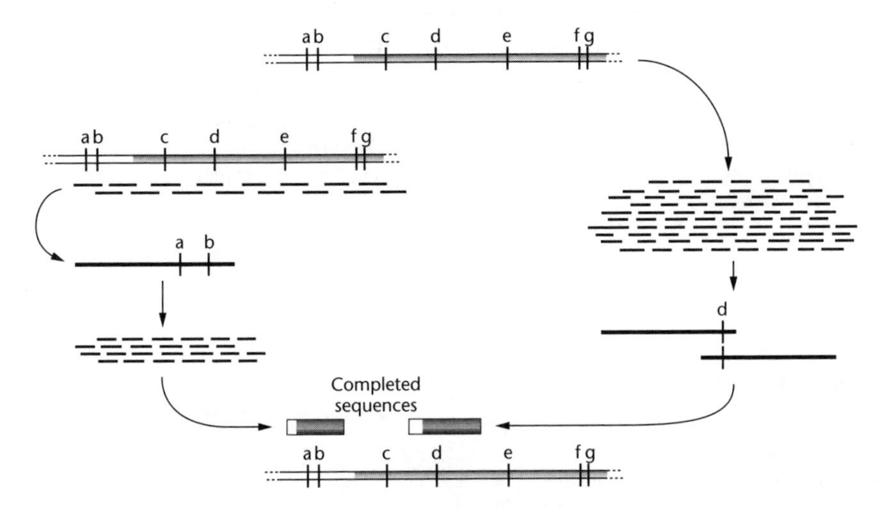

Fig. 2.6 Strategies for genome sequencing. The top panel shows a stretch of genomic DNA 2–3 Mb in length with seven physical markers, such as sequence tagged sites, shown as vertical lines. In the clone-by-clone approach (shown on the left) the genomic DNA is cloned into BAC vectors which are mapped by identifying overlaps and anchored onto the reference physical map using the markers. A minimal tiling path of BAC inserts is assembled to cover the genomic region. Individual BACs (e.g. the left-hand BAC corresponding to markers a and b) are then shotgun sequenced by breaking randomly into small fragments. The sequence is reassembled by computer and the completed sequence placed on the map. In the whole genome shotgun approach (shown on the right) the genomic DNA is shotgun sequenced and reassembled by computer. For small genomes a reference map is not required but for larger genomes, such as the human genome, it was necessary to use existing map data to help assemble the sequence properly.

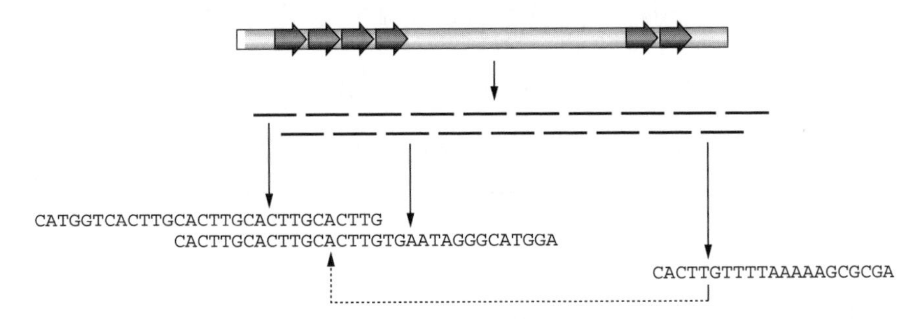

Fig. 2.7 Problems caused by repetitive DNA. The top panel shows a DNA insert from a BAC clone containing dispersed and tandem repeats. When this insert is shotgun sequenced, the repeats can cause mistakes in the alignment. For example, on the left, a false overlap can be generated between two flanking clones, eliminating the two internal repeats. In the case of dispersed repeats, false overlaps can also eliminate unique sequence DNA, which may include genes.

Box 2.6 Draft sequences and finished sequences

The publications in February 2001 announcing the completion of draft human genome sequences were heralded as a great scientific achievement. However, neither sequence covered more than 90% of the genome and much of the data was unrefined. What remains to be done in order to turn a draft sequence into a finished sequence?

Heterochromatin

Most of the missing sequence represents **heterochromatin**, densely packed DNA primarily from the centromeric regions of each chromosome. This comprises huge blocks of tandem repeats that are notoriously difficult to clone. It is possible that some heteromeric DNA will be forever beyond our reach and the human genome sequence will never be absolutely complete. However, since very few genes are found in heterochromatin, this is unlikely to impact on the medical applications of the genome.

Gaps

Gaps arise in all sequencing projects due to **sampling errors**. This is rather like having 100 marbles in a bag and picking them randomly in an effort to pick each one at least once. There are always one or two marbles that "escape" while others may be picked several times. Sampling errors occur during library construction (some parts of the genome are not represented in the libraries) and during the sequencing phase (some sequences just happen not to be picked). Gap closing strategies include using multiple genomic libraries and amplifying genomic DNA with PCR primers facing outwards from the ends of known contigs. There were approximately 50,000 gaps in the draft genome sequences.

Unfinished sequence

Automated sequencing produces data in the form of a **sequence trace**, which is a series of peaks representing different bases (Fig. B2.6). In order to avoid inevitable errors, each part of the genome is independently sequenced 8–10 times before it is said to be **finished**. The quality of the sequence is assessed using a computer program such as PHRED which assigns a score to each peak. If the sequence is judged to be of low quality, then it is rejected and must be carried out again. In both the HGP and Celera outputs, only about 25% of the draft sequences were of finished quality.

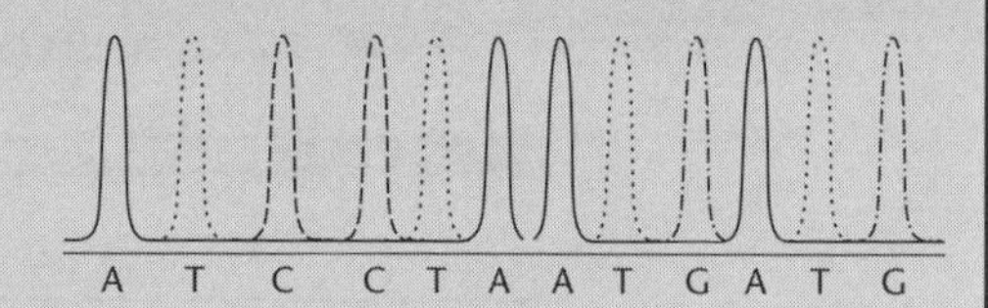

Fig. B2.6 Sequence trace data.

- There is evidence a sequence is expressed.
- A sequence is homologous to a known gene or EST (either human or from another species).
- A sequence carries the hallmarks of a gene, e.g. a promoter, splice sites, a polyadenylation site or a base composition that suggests the existence of an exon.

Computer algorithms are used to search for genes both *ab initio* (looking for gene-like features from first principles) and on the basis of homology. This provides scope for both the overestimation and underestimation of gene number. For example, genes can be falsely predicted if a sequence shows strong homology to a known gene but is in fact a **pseudogene** (a nonfunctional gene relic) or if the prediction is based on a cDNA sequence that is an artifact (genomic sequences can occasionally be incorporated into cDNA clones during library construction). On the other hand, genuine genes can be missed if they are expressed at very low levels or in restricted cell populations, since they will rarely appear in cDNA libraries, or if the features of the gene are not recognized by the computer (this applies especially to atypical genes, such as those for noncoding RNAs). Human genes can be difficult to recognize because they are often very large, but are divided into a number of tiny exons separated by immense introns. Therefore, even if a gene is identified, it is common for exons to be missed or for the boundaries of the gene to be determined incorrectly. There are even examples of small human genes being hidden within the introns of larger ones. It may take a very long time to produce a complete and accurate human gene catalog.

A significant challenge in the latter stages of sequencing was the presentation of sequence data, and the associated gene annotations, in a readily accessible manner. This problem has been largely overcome by the development of **genome browsers** which use front-end graphical user interfaces to present information and allow the user to navigate between screens showing the genome at different resolutions. For example, the EnsEMBL browser (http://www.ensembl.org/) allows the user to view the entire genome as a series of chromosomes. By clicking on a chromosome, the user can home in on a particular subchromosomal region and increase the resolution stepwise right down to the single nucleotide level (Fig. 2.8). Each chromosome segment is extensively annotated with genes, markers and other features which are regularly updated as more information becomes available. There are extensive links to external databases that give further information on the structure and function of genes, and to related genes in the genomes of other organisms.

The future: functional genomics

We now know that there is somewhere in the region of 30,000 human genes. The next task, put simply, is to find out what they all do. We know that our inherited diseases are caused by malfunctioning genes, that genes underlie our responses to drugs, pathogens, and other agents in the environment, and that genes influence our susceptibility to diseases such as asthma which have a significant environmental component. Using conventional methods, however, the gene–disease or gene–response relationship has been unravelled for only about 1500 genes and in each case the path to enlightenment has been slow and laborious.

The aim of functional genomics is to determine the functions of genes on a large scale using new, high-throughput technologies. These technologies therefore represent the new tools of medical discovery. The overall goal is to learn the precise relationships between our genes, or more accurately their protein products, in coordinating the activities of a healthy body. When these activities break down, we

Fig. 2.8 Screenshot from the EnsEMBL human genome browser showing the whole of chromosome 5, an overview of band p13.2, and part of the detailed view of a 100-kb region from within this band. The viewer can scroll down the page and read off the DNA sequence a three-frame translations of both strands.

need to understand what has happened at the molecular level, as this will enable us to develop and implement more effective therapies. The same principles apply to the genes and proteins of our pathogens. The more we understand about them, and how they interact with the proteins in our own bodies, the more we can do to intervene and limit the impact of infectious disease. The technology platforms in functional genomics fall into several key areas whose principles and applications are discussed below.

Table 2.3 Primary and secondary sequence and structural databases.

Database	URL
Primary sequence databases	
GenBank	http://www.ncbi.nlm.nih.gov/Genbank/GenbankSearch.html
EMBL nucleotide sequence database	http://www.ebi.ac.uk/embl/
DNA database of Japan (DDBJ)	http://www.ddbj.nig.ac.jp/
SWISS-PROT and TrEMBL	http://us.expasy.org/sprot/
Secondary sequence databases	
ProSite	http://us.expasy.org/prosite
BLOCKS	http://www.blocks.fhcrc.org
PRINTS	http://bioinf.man.ac.uk/dbbrowser/PRINTS
Pfam	http://www.sanger.ac.uk/Software/Pfam
InterPro	http://www.ebi.ac.uk/interpro/
Protein structure databases	
Protein Databank (PDB)	http://rscb.org
European Macromolecular Structure Database (EMSD)	http://www.ebi.ac.uk/Databases/structure.html

Sequence comparison and comparative genomics

It is possible to find out a large amount of information about the function of a gene without doing any experiments. The computer programs that are used to search for genes in genomic DNA often incorporate algorithms that attempt to find sequences matching known genes by applying the principle of **similarity searching**. These algorithms rely on **sequence databases**, which are universal repositories for sequence information. The importance of databases in genomics cannot be understated. Databases are electronic repositories for all kinds of biologic information, and many of them are freely accessible over the Internet. The **primary databases** store original nucleic acid and protein sequence data while so-called **secondary databases** use the information in the primary databases to compile profiles of highly conserved protein families (Table 2.3). Search algorithms such as BLAST and FASTA and their derivatives allow new sequences to be compared with all the stored sequences in an attempt to find matches. Importantly, the sequence databases do not contain sequences alone but also information that may relate to gene function. If the function of a human gene is unknown, it is often the case that a related gene has been studied in another species and some information about its function is known. Therefore, the quickest way to establish the function of a new gene is often to search the databases and attempt to find related sequences that have already been annotated.

Functional annotation by sequence comparison can be applied on a whole genome scale, but this is not sufficient to determine the functions of all genes. For example, when the yeast genome was sequenced in 1996, 30% of the genes were already known and functions had been assigned by actual experiments. A further 30% of the predicted genes could be assigned tentative functions on the basis of homology to other genes already in databases. This left 30% with no known function (these were called **orphan genes**), while the remaining 10% were regarded as questionable predictions. The functional annotation of the human genome was complicated by the greater difficulty in predicting genes accurately, but overall about 60% of the genes could be assigned a function either by pre-existing experimental evidence or on the basis of homology to human genes or genes from other species that had already been characterized. This leaves 40% of our genes with no known function.

Part of the problem with orphan genes reflects difficulty in the detection of related sequences. The standard search algorithms can reliably detect protein sequences that are 30% identical, but below this level then many relationships are not found. More sophisticated search techniques based on patterns and profiles compiled from large protein families can help in some circumstances but these methods are not always reliable. A relatively new approach is to use protein structures rather than sequences for functional annotation. Protein structures are much more highly conserved than sequences. For example, the structures of hemoglobin and myoglobin are essentially identical even though the sequences show only 17% identity. It is possible to match protein structures showing less than 10% identity with confidence, using structural comparison algorithms such as DALI, VAST, and COMPARER. As for sequences, there are primary databases of protein structures that have been solved by X-ray crystallography and/or nuclear magnetic resonance spectroscopy. Several international collaborative efforts are ongoing to solve protein structures in a systematic and high-throughput manner, so that examples of every protein family in existence are stored in the databases. In the future, it may be possible to identify the functions of orphan genes by expressing them as proteins, solving the protein structures, and comparing these to structures in the databases. If this is the case, however, a standard language for the description of protein structures must be established (Box 2.7).

Another sequence-based method for functional annotation is **comparative genomics**. The principle of comparative genomics is that similarities between species extend far beyond the level of individual genes, and include whole genomes. This can be useful in a number of ways. In closely related species (e.g. mouse vs. human) comparative genomics can help to identify genes and their regulatory elements, since only sequences with an evolutionarily conserved function would be found in both genomes while others would have diverged significantly. There is also a surprising degree of conservation at the level of whole molecular pathways and networks even between distantly related organisms. One striking example is the conservation of the insulin signaling pathway between humans and the nematode *Caenorhabditis elegans*. Indeed, nearly 60% of the known human disease genes have counterparts in this worm and/or in the other invertebrate model organism, the fruit fly *Drosophila melanogaster*. This allows experiments to be carried out in these amenable organisms to identify further pathway components, which can help to isolate corresponding human genes. There are even relatives of about 30% of our disease genes in yeast, representing highly conserved functions such as cell division and DNA repair. As well as providing tools for functional annotation, these organisms can also be used as disease models (see Chapter 8). For example,

Box 2.7 Standardized structural and functional classification of proteins

The functional annotation of genes and their products is often dependent on experiments carried out on different organisms in different laboratories all over the world. Scientists are used to describing essentially the same protein by several different names but this is only a triviality. What is more important, as we move towards a world of increasing data availability and international collaboration, is the need for standardization in the description of protein structures and functions since these relate to precise physical properties.

Several different hierarchical systems have been developed for the classification of protein structures including SCOP (Structural Classification of Proteins), CATH (Class, Architecture, Topology, Homologous superfamily) and FSSP (Fold classification from Structure–Structure alignment of Proteins). The different systems use different criteria and thresholds to recognize **fold groups** (a fold is another way of describing the tertiary or three-dimensional structure of a protein). Therefore, while proteins are treated similarity in the upper levels of the hierarchy, differences start to appear in the more detailed classifications lower down. Additional problems are caused by **superfolds** (structures appearing in several protein families that are only distantly related) and continuous variation between one fold type and another.

The systematic functional classification of proteins has been implemented by the **Gene Ontology Consortium**, which is developing a common system of classification applicable to all organisms (http://www.geneontology.org/). Central to the classification scheme are three separate ontologies to classify proteins according to molecular/biochemical function (e.g. kinase), cellular component (e.g. a particular signaling pathway), and biologic process (which in the case of humans would include the role of the protein in disease). The Gene Ontology system is not restricted to hierarchical classification and is therefore more flexible than any of the current systems for structural classification.

the insulin signaling pathway in *C. elegans* can be disrupted to provide an excellent model of type 2 diabetes.

Although sequence and/or structural comparison can be a useful first recourse for functional annotation, there are several reasons why the information should be treated with caution:

• **Low complexity regions** are found in many proteins with extremely diverse functions. Transmembrane domains fall into this category.

• Sequence similarity does not guarantee functional similarity. A number of sequences are multifunctional and can crop up in proteins with completely unrelated functions. For example, the α/β hydrolase fold forms part of the catalytic site of several functionally distinct enzymes and is also found in a cell adhesion molecule.

• Some proteins have acquired additional functions during evolution. A good example is the recruitment of a variety of rather mundane metabolic enzymes as crystallins, the proteins that allow the lenses in our eyes to refract light.

• The language used by all researchers to describe function is not the same, and ambiguities may arise. Recently, there has been a concerted effort to standardize the nomenclature for functional classification (Box 2.7).

• The databases contain mistakes, some of which are clerical but many of which arise from experimental errors. Annotating a new gene on the basis of database information alone risks endorsing mistakes made by other people.

Finally, most functional predictions based on sequence similarity are predictions of biochemical function, and this may not always be helpful. For example, it may be possible to establish that a new gene encodes a protein kinase, but this does not provide any information about the wider role of the protein. Further information is required about the cellular or biologic level functions of the protein including its role in disease, and this can only be established by other types of experiment.

Transcriptomics: global analysis of mRNA

Certain aspects of gene function can be determined by the analysis of expression profiles. For example, if a gene is expressed in response to growth factor signaling, it can be asserted, with reasonable confidence, that the function of the gene has something to do with cell proliferation. Genes that are expressed in the disease state but not in healthy tissue, or vice versa, are likely to be involved in some way in the pathogenesis of the disease or the body's response to it. Similarly, genes expressed specifically in response to the administration of a drug may hold the key to the basis of an adverse drug reaction. Traditionally gene expression profiles have been studied individually or in small groups, which has limited the scope of investigations to specific candidate genes whose role in disease was already suspected. Since 1995, however, there has been an increasing trend towards the **global analysis of gene expression**, in which the expression profiles of thousands of genes are monitored simultaneously. Ultimately, it will be possible to monitor the entire **transcriptome**, i.e. all the mRNAs in the cell. This type of comprehensive analysis could help to identify all the genes involved in any medically relevant process. Already, transcriptome analysis has helped to identify new disease markers, drug targets, and potentially therapeutic gene products, and has generated data allowing genes to be linked into functional groups. Two technology platforms have been at the forefront of this shift in perspective: sequence sampling and microarrays.

Sequence sampling

Global gene expression profiling by **sequence sampling** is based on a simple principle: strongly expressed genes will produce more mRNA than weakly expressed genes and therefore will be better represented in cDNA libraries. The partial sequencing of thousands of randomly picked cDNA clones followed by counting the number of times each gene is represented therefore provides a rough guide to the relative levels of gene expression. Furthermore, comparing two cDNA libraries (e.g. one derived from healthy tissue and one derived from a clinical sample such as a tumor) can show which genes are differentially expressed, and therefore which ones are altered in the disease state.

Although straightforward, sequence sampling is laborious and expensive, since each experiment involves a large scale sequencing endeavor. To overcome this limitation, several high-throughput sequence sampling techniques have been developed and the most established of these is **serial analysis of gene expression (SAGE)**. The SAGE procedure is outlined in Fig. 2.9. The details are complex, but

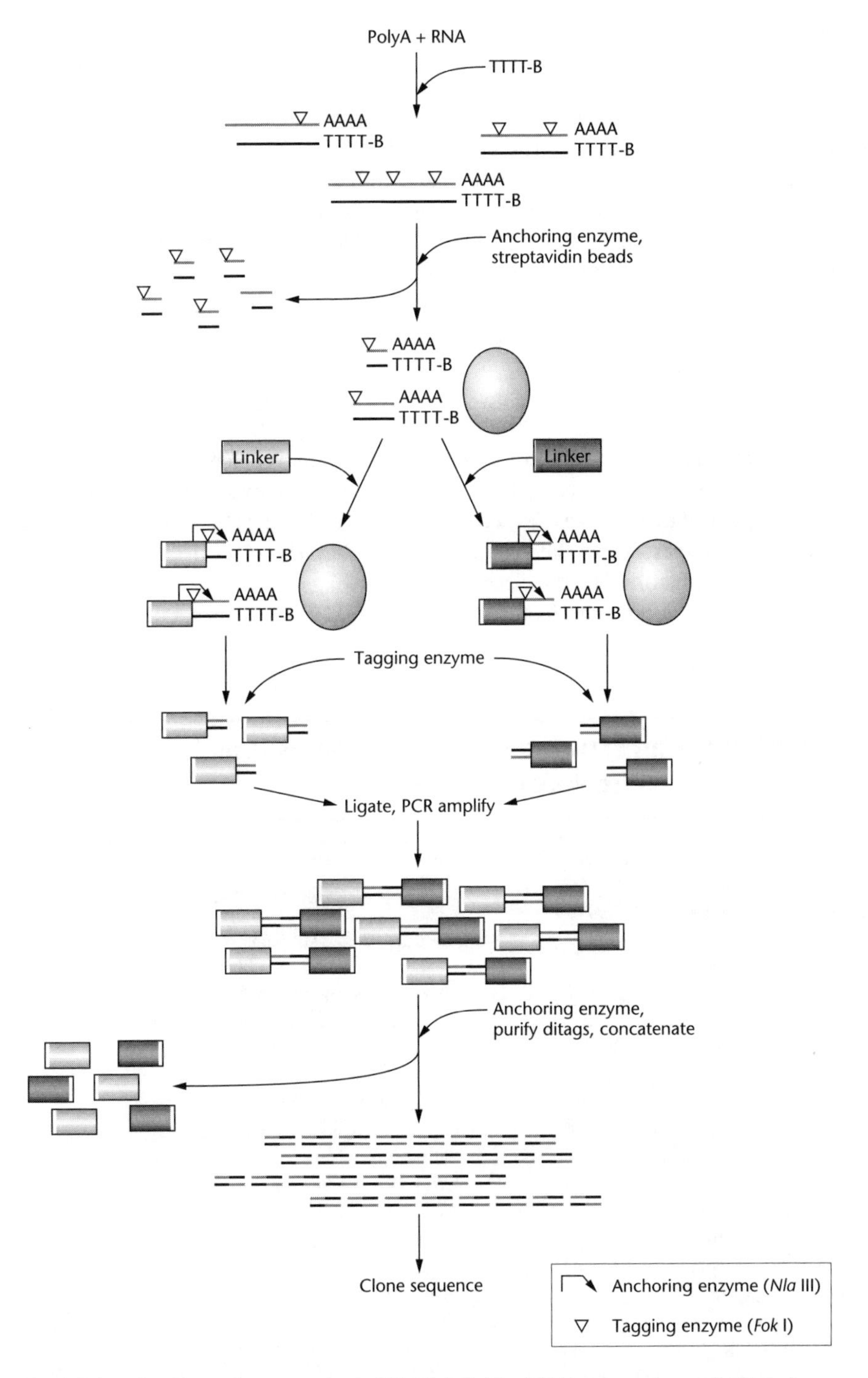

Fig. 2.9 Serial analysis of gene expression (SAGE). PolyA$^+$ mRNA is reverse transcribed into cDNA using an oligo-dT primer conjugated with biotin. A frequently cutting restriction enzyme, the anchoring enzyme, then cleaves the cDNA to leave short sequences attached to biotin, which are rescued with streptavidin beads. Linkers are added to these sequences and the linkers contain the recognition site for a special restriction enzyme, the tagging enzyme, which cleaves a certain distance downstream from the recognition site. Cleavage with the tagging enzyme generates very short tags attached to the linkers. Ligation and PCR amplification of the linkers generates ditags which can be isolated from the linkers with the original anchoring enzyme and joined into long concatemers for cloning. Each tag identifies a gene, and the number of times each tag appears provides a guide to that gene's expression level.

again the principle is simple. Instead of sampling thousands of cDNAs individually, restriction enzymes are used to prepare very short sequence tags (8–15 nucleotides) from each cDNA and these tags are joined together to form long concatemers. The concatemers are sequenced, and since each contains tags representing 50–100 cDNAs, the amount of data that can be collected from one experiment is increased dramatically. As with cDNA sampling, the tags are counted and this provides a guide to relative gene expression levels. Databases have been set up to collect SAGE data and allow comparisons across different experiments.

Microarrays

DNA microarrays are miniature devices containing many different DNA sequences, or **features**, each one representing a different gene. The two major types of array are the **spotted microarray**, which is made by transferring DNA sequences onto a coated glass microscope slide, and the **high-density oligonucleotide chip**, which is manufactured by the direct synthesis of oligonucleotides on a glass wafer (Box 2.8). Although manufactured in completely different ways, the principles

Box 2.8 Manufacturing spotted microarrays and oligonucleotide chips

The first DNA arrays were prepared by transferring DNA clones to nylon filters by hand. They were large and cumbersome, and relied on the use of radioactive probes. In order to incorporate enough features for whole genome analysis, extensive miniaturization of the array format was required. This was achieved by the adoption of glass as the array substrate. Unlike nylon, glass has a low autofluorescence and therefore permits the use of high-resolution fluorescent probes. Consequently, features can be packed much more closely together (0.3–0.5 mm spacing) allowing microarrays with a density of 5000 features per cm^2 to be produced on a routine basis. Initially, prefabricated cDNA microarrays were prohibitively expensive and were beyond the reach of all but the best-funded laboratories. In recent years, however, increased competition between suppliers combined with the establishment of in-house production facilities at many universities and other institutions has forced prices down.

Arrays of oligonucleotides can be manufactured in the same manner as cDNA arrays, i.e. by transfer to specific positions on a glass slide using a precision robot. However, greatly increased density can be achieved by *in situ* oligonucleotide synthesis using a photolithographic process developed by Steve Fodor and colleagues at the US company Affymetrix Inc. Commercially available Affymetrix GeneChips have a feature density of 64,000 units per cm^2, while even greater densities (250,000–1,000,000 features per cm^2) have been achieved in experimental versions. The first step in the production process is to coat the glass substrate in such a way that DNA can be attached covalently to the surface. Oligonucleotides are then built up at specific positions by the stepwise addition of particular deoxynucleotide triphosphates (dNTPs). Each dNTP is modified to contain a photolabile protecting group so that the DNA molecule cannot be extended unless it is activated by light. The application to the chip of a series of photolithographic masks marks out those areas where the DNA can be extended and those areas where the DNA remains inert. In this way, specific sequences can be built up at predetermined sites. Because the technology is owned and licensed exclusively by Affymetrix Inc. it remains very expensive both in terms of the chips themselves and the equipment and software needed to use them. In many cases it is less expensive to contract out the experiments to dedicated laboratories.

involved in global gene expression analysis are much the same for each device. In both cases, expression profiling is based on **multiplex hybridization** using a complex population of labeled DNA or RNA molecules as the hybridization probe.

The typical experimental procedure used with spotted arrays is shown in Fig. 2.10. To prepare the complex probe, mRNA from a particular source (e.g. healthy human liver) is converted into cDNA using a reaction mixture containing fluorescent nucleotide analogs. The cDNA population is therefore **universally labeled**, i.e. every single cDNA molecule contains fluorescent nucleotides. This complex mixture of cDNA molecules is representative of the original mRNA population, i.e. there will be many cDNA copies of the abundant transcripts and few copies of the rare transcripts. This probe mixture is applied to the array and individual molecules hybridize to their complementary targets. On spotted arrays, each feature or spot contains 10^6–10^7 copies of the same DNA sequence, which far exceeds the number of copies of any particular cDNA in the probe even if that cDNA represents an abundant transcript. Under these conditions, **nonsaturating hybridization** will occur and the intensity of the hybridizing signal at each spot on the array will be proportional to the relative abundance of that particular cDNA in the probe. Therefore, the relative expression levels of thousands of different transcripts can be monitored in one experiment.

The comparison of different samples (e.g. healthy liver vs. hepatoma) can help to identify differentially expressed genes. Such comparisons can be achieved by hybridizing complex probes from each sample to copies of the same array. However, more direct visualization can be achieved by labeling the two cDNA populations with different fluorophores and hybridizing them to the same array. The ratio of the signals from each fluorophore provides a comparison of expression levels in the two samples. Typically, the array is scanned at two emission wavelengths and a computer is used to combine the images and render them in false color. If one fluorophore is portrayed as green and the other as red, features representing differentially expressed genes will show up as either green or red, while those genes expressed at similar levels in each sample will show up as yellow (Fig. 2.11).

Applications of transcriptomics

Sequence sampling and SAGE have been used to characterize gene expression profiles associated with a number of diseases, including athersclerosis, HIV infection, and cancer. Spotted arrays and GeneChips have also been used extensively for profiling diseases, including asthma, rheumatoid arthritis, athersclerosis, diabetes, inflammatory bowel disease, cancer, and cytomegalovirus infections. In all these studies, large numbers of genes were shown to undergo changes in expression levels providing evidence that disease and environmental change result in large scale changes to the transcriptome. In some cases, such experiments have helped to identify new disease markers. For example, the *CD36* gene has been identified as a novel marker for insulin resistance. Other studies have provided useful potential drug targets. For example, the *EGR1* gene was shown to be upregulated fivefold in athersclerosis, a disease in which cholesterol-rich white blood cells are deposited on the inner surface of arteries. The product of the *EGR1* gene is a transcription factor with many

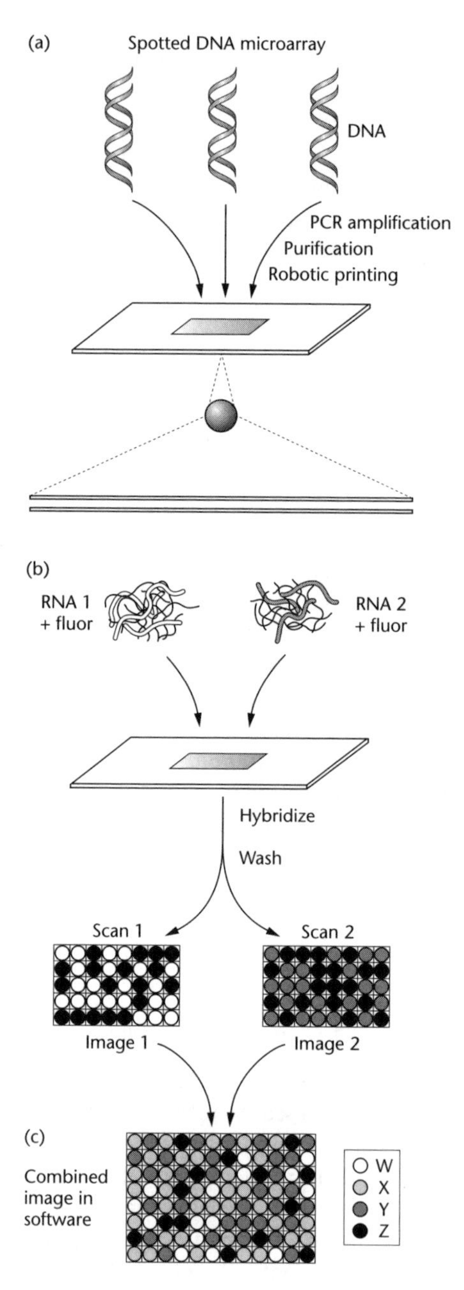

Fig. 2.10 Spotted DNA microarrays. (a) Principle of manufacture (see Box 2.8 for details). (b) Comparative analysis of gene expression levels (see text for details). (c) Combined image showing four types of signal: W, genes expressed equally in each mRNA sample; X, genes expressed more strongly in sample 1; Y, genes expressed more strongly in sample 2; Z, genes expressed in neither sample.

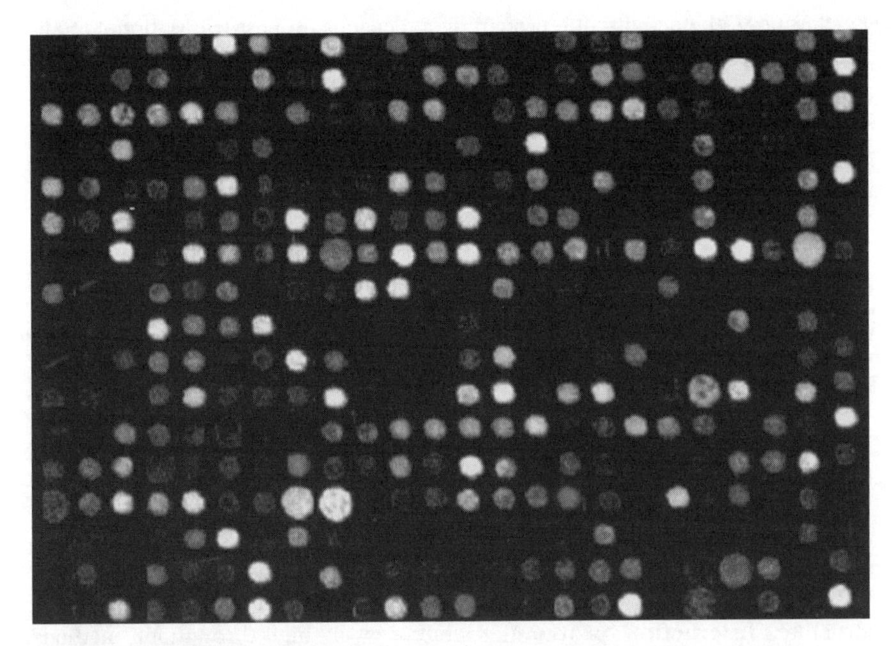

Fig. 2.11 Raw data from a microarray experiment shown as a monochromatic image. Signal intensities at different addresses on the array relate to different steady state mRNA levels in the sample. For comparison of mRNA levels over two experimental conditions, two fluorophores with different excitation and emission wavelengths are used. The output from such experiments can be rendered in false color.

known target genes, some encoding proteins involved in blood coagulation and cell adhesion with likely roles in disease pathogenesis. In other cases, the studies have provided novel insights into biologic processes. For example, the addition of serum to quiescent fibroblasts in culture resulted in the expected induction of proliferation response genes but also of genes involved in wound healing, such as *VEGF*, which promotes the growth of new blood vessels. Quite apart from its application as a discovery tool, transcriptomics is increasingly being used for diagnosis. Differences in gene expression profiles can help distinguish between diseases that are very difficult to tell apart using conventional tests. As an example, the related diseases acute myeloid leukemia and acute lymphoblastoid leukemia are discussed on page 120.

Proteomics: global analysis of proteins

The transcriptome clearly provides a useful way to study and characterize diseases and our responses to drugs or changes in the environment. However, in most cases, the actual functional molecules in the cell are not mRNAs but proteins. The abundance of a particular transcript is not always related to the level of the corresponding protein because protein synthesis is often regulated independently.

Furthermore, the activity of a protein often depends on post-translational modifications such as phosphorylation, and these cannot be predicted from mRNA levels. Therefore, to get an accurate picture of the functioning cell in health and disease it is necessary to study the proteins directly. The complete set of proteins found in a given cell is known as the **proteome**, and the study of proteins on a global scale is therefore known as **proteomics**.

While both transcriptomics and proteomics are concerned with monitoring gene expression, proteomics also encompasses several additional areas of investigation. The protein equivalent of transcriptomics is sometimes termed **expression proteomics** and involves large scale approaches that investigate and compare the abundance of proteins in different samples. Many of the aims of expression proteomics are similar to those of transcriptomics – the identification of markers, drug targets, proteins with therapeutic potential, and the use of diagnostic expression profiles – but there is the advantage that changes in protein levels reflect the changing biochemical activity of the cell directly and that proteomic methods can distinguish proteins that are modified, e.g. phosphorylated and nonphosphorylated versions. For example, a protein known as stathmin is a useful marker of childhood leukemia, but only the phosphorylated version is diagnostic of the disease (see p. 121).

The other branches of proteomics concern protein interactions and protein structure. **Interaction proteomics** embraces all high-throughput methods for studying protein–protein interactions and protein interactions with other molecules. The investigation of protein–protein interactions can help to assign proteins into complexes or pathways, and this is useful both in terms of the fundamental knowledge it provides and the potential impact on drug development. For example, if a cell surface receptor is identified whose overactivity is responsible for a particular disease, it might represent a potentially very useful drug target. However, if the gene encoding that receptor shows extensive polymorphic variation in the population, it would be difficult to identify a lead compound suitable for everyone. A drug company might try to identify multiple leads which could be used to treat particular subsets of patients, but a preferable approach would be to identify proteins interacting with the receptor in an attempt to find surrogate targets with less polymorphism. Drugs interacting with the second target would achieve the same aim, i.e. the disruption of the signaling pathway.

By investigating protein interactions with small molecules, suitable lead compounds can be chosen on a rational basis rather than carrying out random screening of chemical libraries (see Chapter 7). Such interactions can be studied using molecules similar to known ligands or molecules that have been shown in computer simulations to complement the structure of the target protein. This **rational drug design** approach was instrumental in the development of drugs such as captopril and zanamivir (Relenza). Interaction modeling requires the availability of solved proteins structure corresponding to drug targets. As discussed above, a number of **structural proteomics** initiatives have been initiated with the grand purpose of solving representative structures of all protein families. However, there are also medically oriented projects whose aim is to solve the structures of proteins involved in human diseases.

Box 2.9 Protein microarrays

Protein microarrays are miniature devices, similar in principle to DNA microarrays, which are used to analyze proteins. They are manufactured in much the same way, by the robotic spotting of protein samples onto treated glass slides. The two major types of array are **analytic arrays** and **functional arrays**. Functional arrays are used to investigate the biochemical activities and interactions of proteins, so the proteins under investigation are themselves immobilized on the array. By contrast, analytic arrays are used for expression monitoring, and the proteins under investigation are in the sample applied to the array. The array itself contains specific **capture agents**, which may themselves be proteins (e.g. antibodies, lectins) or may be other types of molecules (e.g. oligonucleotide aptamers). The use of lectin arrays to study glycoproteins is discussed in more detail in Box 6.5. There are also other types of analytic devices known as **protein biochips**. These do not capture specific proteins, but they have different surface chemistries that bind particular classes of proteins.

Protein microarrays have the same advantages as DNA arrays, i.e. they allow the parallel analysis of many proteins at the same time in a miniature and automated assay format. They can be integrated easily with downstream mass spectrometry analysis, and they are amenable to novel detection methods such as **surface plasmon resonance spectroscopy**, which means that the proteins do not have to be labeled. A possible disadvantage is that proteins are chemically very diverse and it is difficult to design a universal set of conditions that will suit all the proteins on the array. Even so, a functional array carrying most of the 6000 or so yeast proteins has recently been produced which should make the analysis of the yeast proteome much more straightforward. The human proteome is an order of magnitude more complex than the yeast proteome due to alternative splicing and post-translational modification, so a human proteome array is unlikely to be developed in the near future.

Technology platforms for proteome separation

The proteome of a typical human cell is extremely complex, containing up to 100,000 different molecules differing in abundance over four or five orders of magnitude. Unlike nucleic acids, proteins cannot be amplified using a simple technique like the PCR, and because proteins are physically and chemically extremely diverse, there is no procedure equivalent to hybridization which can be used to assay all proteins at the same time. Therefore, while protein chips have been developed for expression analysis (Box 2.9), the major technology platforms in expression proteomics are based on the separation of complex mixtures and the identification of proteins in individual fractions by mass spectrometry. Currently, the most popular separation method is **two-dimensional gel electrophoresis (2DGE)**.

Because the proteome is so complex, a separation process based on one property (e.g. size, charge, solubility) would not provide sufficient resolution. The advantage of 2DGE is that proteins are separated in one dimension on the basis of charge and then in the second dimension on the basis of mass (Fig. 2.12). The charge separation is achieved by **isoelectric focusing**, where denatured proteins are separated in an electrophoretic strip gel that has a **pH gradient**. When the electric field is applied, the proteins migrate to the position where their net charge is equivalent to the surrounding pH. The gel is run for a long time so that the proteins focus at their **isoelectric points** regardless of their size. The strip gel is then equilibrated with a solution of the detergent **sodium dodecylsulfate (SDS)**, which carries a large

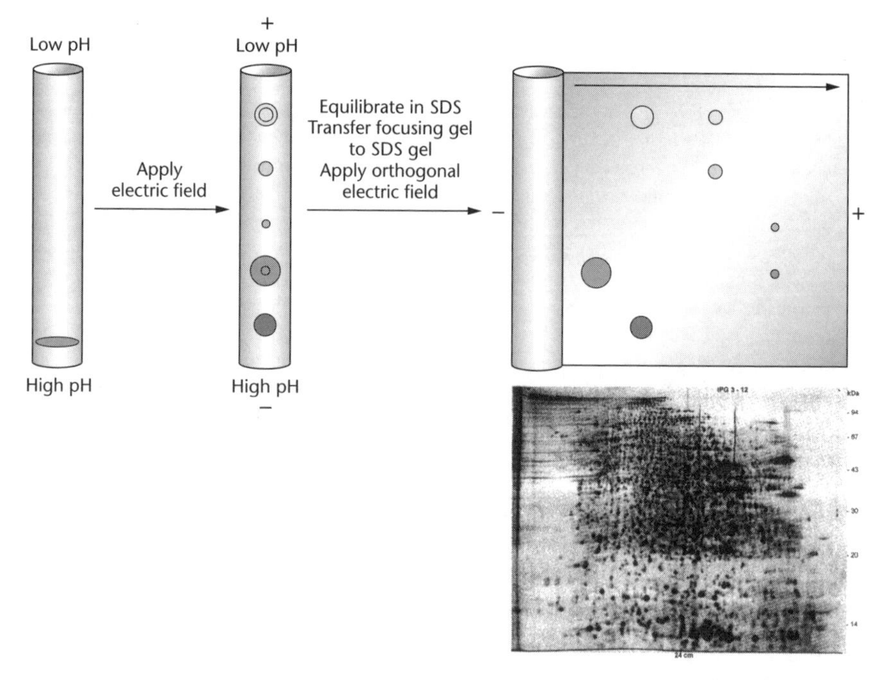

Fig. 2.12 Principle of two-dimensional gel electrophoresis. A denatured protein sample is loaded at one end of an isoelectric focusing (IEF) gel and proteins are separated according to their overall charge, **regardless of their mass**. The IEF gel is then equilibrated in SDS and attached to a standard SDS–PAGE gel for orthogonal separation on the basis of mass, **regardless of charge**. The charge of the protein was represented by different shades of color from red (basic) through to yellow (acidic) while the mass is represented by the sizes of the circles. The inset image shows the typical appearance of protein spots on a two-dimensional gel after staining.

negative charge and binds to the backbone of denatured proteins stoichiometrically, effectively canceling out any intrinsic charge each protein may possess. The strip gel is placed at one edge of a polyacrylamide slab gel and the electric field reoriented by 90 degrees. Size separation is achieved by standard **SDS-PAGE**, with proteins moving towards the anode through the pores of the gel, smaller proteins passing more easily through the pores than larger ones and therefore moving further. The gel is usually post-stained and the proteins are revealed as a pattern of spots (see inset to Fig. 2.12).

Owing to certain limitations of 2DGE (e.g. poor sensitivity for proteins with very low abundance, poor representation of certain protein classes including membrane proteins) **multidimensional chromatography** is becoming more popular as an alternative separation method. Any method that partitions a protein mixture between a solid stationary phase and a liquid mobile phase is known as **chromatography**. The stationary phase (or matrix) is generally supported on some form of column and the mobile phase flows through it. The chromatography method of choice in proteomics is **high performance liquid chromatography (HPLC)**, which involves forcing the mobile phase through the column under high pressure. One of the main

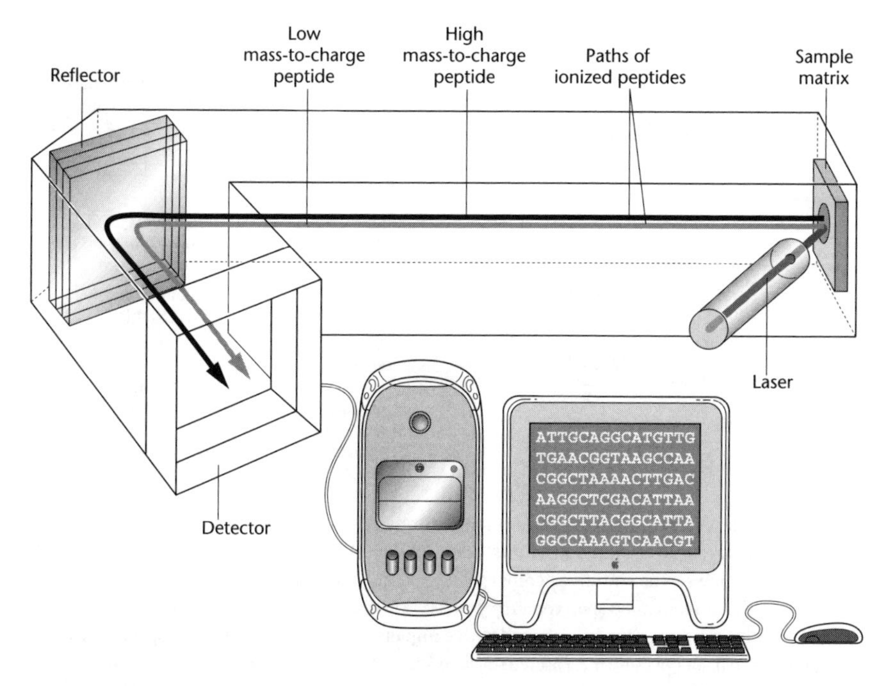

Fig. 2.13 MALDI-TOF mass spectrometry. The sample (or analyte) is embedded in a matrix compound which can absorb laser energy and release it as heat. A short laser pulse causes the sample and matrix to vaporize and emit gas phase ions, which travel down a flight tube and are reflected onto an ion detector. The time of flight is related to the mass/charge ratio of the ions, from which the masses can be accurately determined.

advantages of HPLC, apart from its speed and sensitivity, is the ease with which it is automated and integrated with downstream analysis by mass spectrometry.

Protein characterization by mass spectrometry

The protein spots on 2D gels or the fractions eluting from HPLC columns are **anonymous**. Until recently, the identification of anonymous proteins required either the use of a specific capture agent, generally an antibody, or the sequencing of the protein by stepwise Edman degradation. Neither method is suitable for proteome scale analysis, where thousands of proteins must be characterized rapidly. In the early 1990s it was realized that rapid protein identification could be achieved using emerging mass spectrometry techniques. **Mass spectrometry (MS)** is used to determine the accurate masses of molecules, but for a long time the technique could not be used with large molecules, including proteins, because the ionization process led to fragmentation of the molecule. The development of soft-ionization techniques such as **matrix-assisted laser desorption/ionization (MALDI)** and **electrospray ionization (ESI)** made the mass determination of large molecules possible (Fig. 2.13). In 1993, several independent research groups

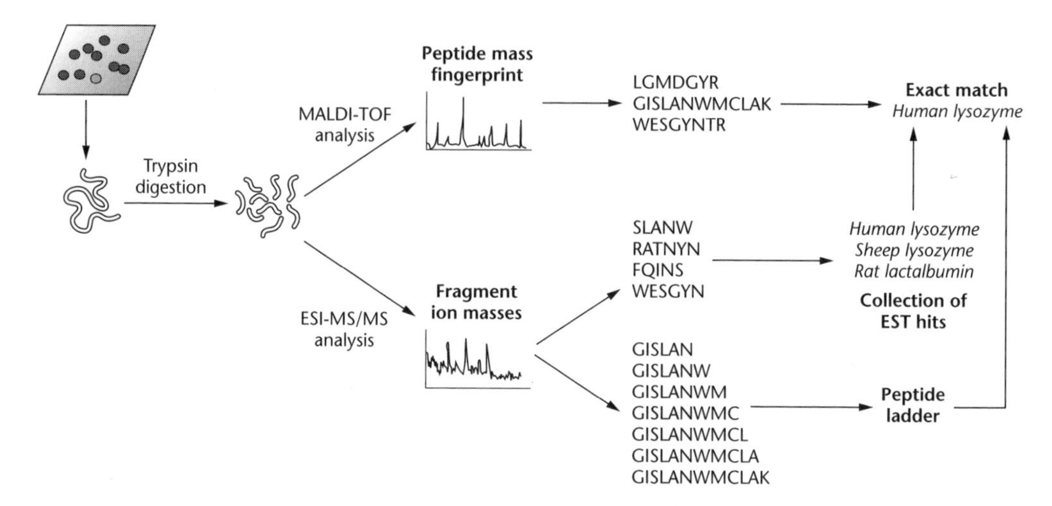

Fig. 2.14 Protein annotation by mass spectrometry. A protein sample from a two-dimensional gel or other separation method is digested with trypsin and placed in a mass spectrometer. MALDI-TOF analysis determines the masses of intact tryptic peptides, which can be searched against protein databases for matching peptides (peptide mass fingerprinting). Alternatively, tandem mass spectrometry (ESI-MS/MS) can be used to determine the masses of fragment ions from each peptide. Databases can be searched with completely uninterpreted MS/MS spectra, with short peptide sequence tags derived from partially interpreted mass spectra, or with longer sequences derived *de novo* from fully interpreted mass spectra.

published algorithms that could be used for correlative searches of protein sequence databases using mass spectrometry data. The principles of this technique, known as **peptide mass fingerprinting**, are as follows:

- A protein sample, e.g. from a 2D gel, is digested with the protease trypsin to produce a collection of **tryptic peptides**.
- These are analyzed by MALDI MS, usually with a time of flight (TOF) detector.
- The masses of the peptides are determined.
- A search algorithm such as MS-BLAST uses the peptide masses to search protein databases. The algorithm carries out **virtual trypsin digests** of all the proteins in the database and calculates the masses of the predicted tryptic peptides. It then attempts to match these predicted masses against the experimentally determined ones.
- A list of matches is returned. If several peptides match the same protein in the database, it is reasonably certain that a complete match has been obtained. All the peptides may not match due to unanticipated post-translational modifications, splice variants, and polymorphisms.

Where matches cannot be obtained, more aggressive tandem mass spectroscopy (MS/MS) techniques can be used to fragment the peptides and calculate the masses of the fragment ions. These can be searched against less robust sequence resources, including EST databases, to obtain partial matches. The fragments can also be assembled into a peptide ladder whose masses can be compared to standard tables of amino acids in order to determine the sequence *de novo* (Fig. 2.14).

Applications of expression proteomics

Expression proteomics, like transcriptomics, is used to identify new disease markers or proteins whose abundance correlates to particular changes in the environment (e.g. the presence of drugs or toxins). The identification of differentially expressed proteins can be achieved in the same way as differentially expressed transcripts, i.e. multiple 2D gels can be compared to look for quantitatively distinct spots, or the proteins in two samples can be labeled with different fluorophores prior to separation and proteins enriched or depleted in one sample identified by the characteristic emission spectrum. This latter method is known as **difference gel electrophoresis (DIGE)**. An alternative procedure which is applicable to HPLC separation is to use **isotope-coded affinity tags (ICATs)**. These are labels carrying different isotopes that can be distinguished by a mass spectrometer. The heavy and light labels are applied to the different samples (e.g. healthy and disease tissue) and then the samples are mixed prior to protein isolation so that purification losses are equivalent. The proteins are separated by 2DGE or HPLC and the fractions analyzed by mass spectrometry. The relative amounts of the two isotopes are measured in each fraction and a significant difference indicates that the protein in that fraction is either more abundant or less abundant in the disease tissue. This protein can then be investigated as a potential disease marker or drug target.

Differential protein expression has been used to identify a number of useful markers and potential drug targets including proliferating cell nuclear antigen (upregulated in breast cancer) and cyclo-oxygenase 2 (downregulated in colorectal cancer). As stated above, a protein called stathmin (in its phosphorylated form) has been shown to be upregulated in childhood leukemia. Several markers of adverse drug responses have also been identified, including a striking fall in the abundance of a calcium-binding protein called calbindin in response to the immunosuppressant drug cyclosporin A. This drug, which is used to prevent organ transplant rejection in children, has nephrotoxic side effects involving calcification of the kidney tubules. The loss of calbindin provides evidence for the molecular basis of these adverse effect and indicates that restoring the level of calbindin could help to prevent them.

Technology platforms for interaction proteomics

Methods such as affinity chromatography and coimmunoprecipitation have been used for many years to study the interactions of individual proteins. More recently, they have been used to identify systematically the components of protein complexes including the human spliceosome, nuclear pore complex, and anaphase-promoting complex (which may have a role in preventing cancer). The complexes are isolated using antibodies that recognize one of the proteins, so cells must be lysed gently to preserve the interactions within the complex. The complexes may then be analyzed directly by mass spectrometry or may first be separated into individual components by electrophoresis (Fig. 2.15). This approach can define all the proteins that function in a complex and may result in the functional annotation of one or more orphan genes. For example, analysis of the epidermal growth

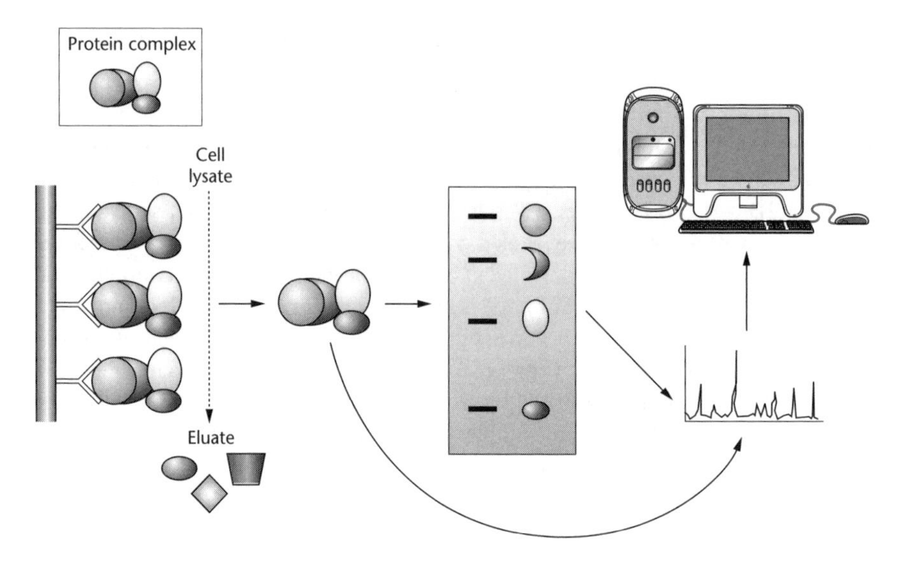

Fig. 2.15 Analysis of protein complexes. A complex with four components can be isolated from a cell lysate by affinity chromatography, in which the cell lysate is passed through a chromatography column containing a matrix onto which is immobilized a capture agent, such as an antibody, which recognizes one of the components of the complex. Other proteins, which are not part of the complex, are washed through. The complex can then be removed from the column and analyzed by mass spectrometry, with or without prior separation by gel electrophoresis.

factor signaling pathway using this method identified nine components, seven of which were already known to be involved, one of which was a known protein but whose role in epidermal growth factor (EGF) signaling was unknown, and one of which was a completely uncharacterized protein. The EGF signaling pathway is implicated in several cancers so this experiment identified two additional potential drug targets.

Interaction screening on a larger scale can be carried out by **phage display** (Fig. 2.16) or the **yeast two-hybrid system** (Fig. 2.17). Each is a library-based system which involves the use of **bait proteins** to trap their interacting partners, which are known as **prey proteins**. In phage display, the bait protein is coated onto a microtiter dish. A phage display library is created by inserting the genes for potential prey proteins into the phage coat protein gene, so that the protein is expressed as a fusion with the coat protein and is expressed on the phage surface. Millions of phage displaying different proteins can then be added to the microtiter dish. After a short incubation, the phage library can be washed away and only those phage displaying proteins that interact with the bait will be retained. These can be eluted and used to infect bacteria, resulting in a large homogenous phage population from which the prey protein gene can be isolated and identified.

The yeast two-hybrid system is an *in vivo* assay in which the bait protein is expressed in yeast as a fusion (a hybrid) with the DNA-binding domain of a transcription factor. All the prey proteins are expressed as fusions with the transcription

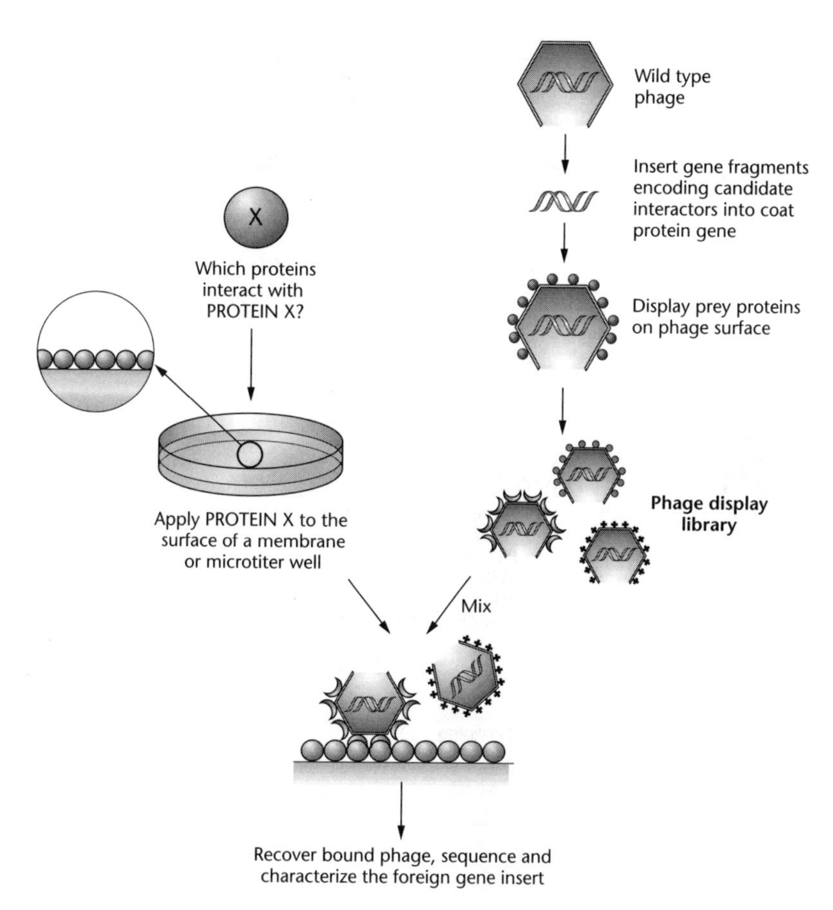

Fig. 2.16 The principle of phage display for protein interaction analysis. The bait protein (X) is attached to the surface of a solid substrate, such as a membrane or a microtiter dish. Bacteriophage expression libraries are then created in which many different proteins are expressed as fusions to the bacteriophage coat protein, resulting in their display on the surface of the phage particle. The phage display library is then added to the microtiter dish, incubated for a short period, and washed away. Only phage displaying proteins that interact with protein X are retained. These can be removed, amplified, and the interacting protein identified by sequencing the foreign insert in the coat protein gene.

factor's transactivation domain. The bait yeast strain is mated either systematically or randomly to a prey library. The products of the mating are diploid cells containing the bait construct and one of the many prey constructs in the library. If the bait protein happens to interact with the prey, a functional transcription factor is assembled that can activate a test gene which is also introduced into the cell. However, if the bait and prey do not interact, the transcription factor remains as two separate hybrid proteins and the gene remains inactive. Large scale interaction screening using the yeast two-hybrid system has yet to be applied to the human proteome, but it has been used to study protein interactions in human pathogens (e.g. hepatitis C virus) and a pilot study has recently been carried out on the mouse proteome.

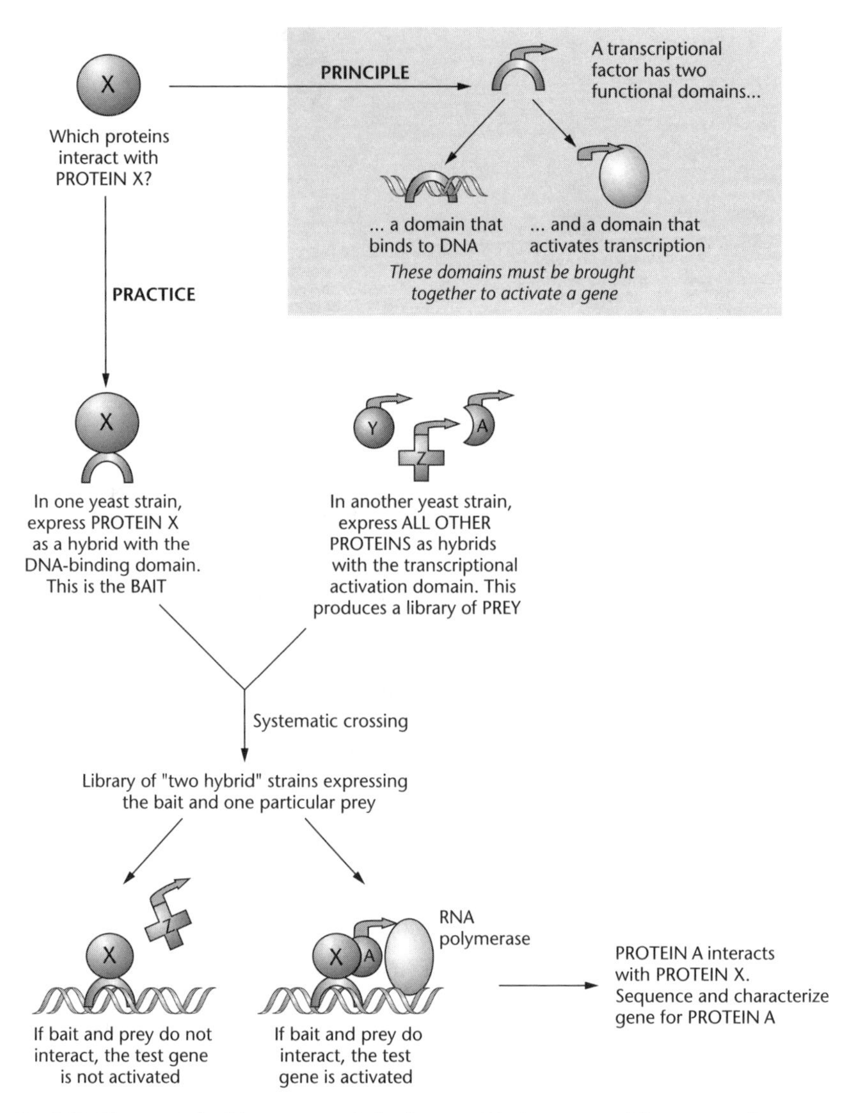

Fig. 2.17 The principle of the yeast two-hybrid system. The bait protein (X) is expressed as a fusion (or hybrid) with the DNA-binding domain of a transcription factor. Yeast expression libraries are then created in which many different proteins are expressed as fusions with the transcription-activating domain of a transcription factor. Mating between bait and prey strains of yeast results in diploid cells expressing both the bait and a candidate prey. Where interactions occur, a functional transcription factor is assembled and a test gene is activated. The interacting protein can be identified by sequencing the expression construct in the corresponding clone in the prey library.

Mutational genomics

All of the functional genomics technologies discussed above help to establish molecular/biochemical or cellular functions. Some of them provide evidence for biologic (whole organism) functions, but none provides a direct link between genes and disease. The direct approach is to mutate genes in model organisms, particularly the mouse, and look for phenotypes that resemble human diseases. As discussed in Chapter 8, disease models have been provided by sporadic mutations and by deliberate gene knockout techniques, but more recently there has been increasing interest in carrying out **whole-genome mutagenesis programs** with the aim of generating **comprehensive mutant libraries** representing hits in every single gene. These can be used as a central resource so that when researchers find evidence from another experimental approach that a particular gene is involved in a disease, the corresponding mutant can be "ordered" and used to study the molecular pathology of the disease and to develop and test new therapies.

Whole genome mutagenesis (**saturation mutagenesis**) is not a new idea. Indeed it has been used for many years to recover mutants affecting particular biologic processes. The difference in the functional genomics approach is that where in the past scientists have discarded mutants that did not interest them, now every mutant is interesting. Recent mutagenesis screens in mice using the potent chemical mutagen ethylnitrosourea (ENU) involved the analysis of some 40,000 lines for medically relevant phenotypes, including allergy, immunology, physiology, developmental defects, behavior, and clinical biochemistry.

In addition to chemical mutagens, further mouse mutagenesis programs have been initiated using **DNA-based mutagens**. These are DNA sequences that are introduced into the genome in a deliberate effort to disrupt genes and cause disease-like phenotypes. The advantage of this approach over chemical mutagenesis (or mutagenesis by irradiation) is that the interrupted gene is tagged with an alien sequence that can be detected by hybridization or PCR, therefore leading to the rapid isolation of the gene and its identification (Fig. 2.18). In this way, the difficult and sometimes obscure pathway from gene to disease becomes a roaring motorway, a rapid conduit for medical discovery in which genes corresponding to particular disease phenotypes can be easily identified.

DNA-based mutagenesis is advantageous for other reasons. By clever modification, the integrating construct becomes able to report certain properties of the interrupted gene, such as its expression pattern. This is achieved by placing a **reporter gene** (a gene with an easily detectable product) inside the construct, but making its activity dependent on the interrupted gene's promoter (Fig. 2.19a). The reporter gene *lacZ* from the bacterium *E. coli* produces an enzyme called β-galactosidase that has the ability to convert a colorless substrate (X-gal) into a blue precipitate. If a construct containing the *lacZ* gene integrates inside a gene that is normally expressed in the pancreas, then the gene is mutated (hopefully giving a disease phenotype) and β-galactosidase is produced in the pancreas, which can be detected by staining tissue sections or whole embryos with X-gal. Since only 3% of the mouse genome is represented by genes, this **gene trapping** approach also helps

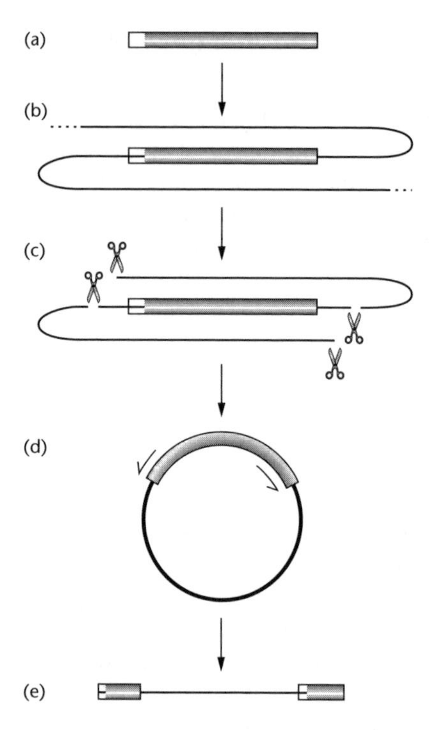

Fig. 2.18 Inverse PCR, one of several methods for the identification of genes interrupted by insertional mutagenesis. (a) The insertion vector. (b) Integration of the vector into genomic DNA. (c) Cleavage of the genomic DNA with restriction enzymes that do not cut within the vector. (d) Circularization of the genomic DNA with DNA ligase, and annealing of PCR primers facing **outwards** from the insertion vector. (e) PCR amplification clones the genomic sequences flanking the insert.

to identify the productive insertions even though the majority within intergenic DNA integrate. A disadvantage of standard gene trapping is that expression of the reporter relies on the expression of the interrupted gene, so if the gene is not expressed the reporter is not expressed either. This problem has been addressed by making the expression of a second gene, this one a selectable marker whose expression is required for cell survival, dependent on the interrupted gene for polyadenylation (Fig. 2.19b). This type of approach has been applied on a large scale, with two major initiatives underway in mice, one organized by the German Gene Trap Consortium (which aims to produce and characterize 20,000 gene trap lines) and one organized by the US company Lexicon Genetics. In both cases, databases of insert flanking sequences are maintained that can be searched by investigators who might find evidence that a particular gene is linked to a disease. Again, the mutant mouse strain can be provided for further experiments.

Mutagenesis programs (chemical and DNA-based) are also ongoing in other model animals, such as *Drosophila* and *Caenorhabditis*. However, in both species an alternative method for functional inactivation is gaining popularity. This method is **RNA interference**, in which double-stranded RNA is processed into

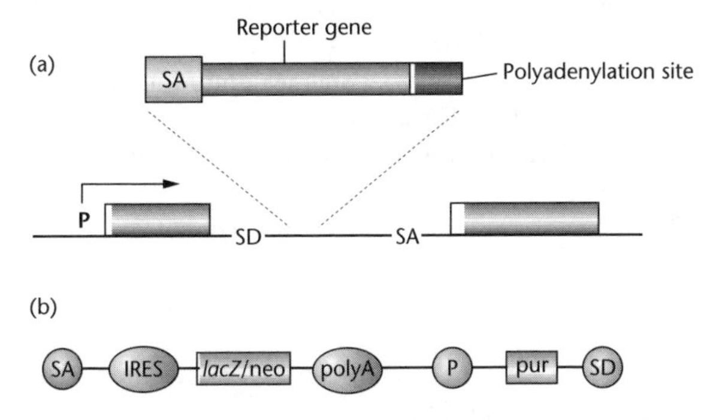

Fig. 2.19 Gene trap vectors. (a) The gene trap contains a splice acceptor site (SA), a reporter gene, and a polyadenylation site. If this integrates inside an existing gene, in either an intron or an exon, it will behave as a terminal exon resulting in a disruption of normal gene function and the expression of a truncated fusion product that has reporter activity and reveals the gene's normal expression pattern. However, this relies on the activity of the endogenous gene's promoter so many genes are missed because of their low-level or restricted expression. (b) A more advanced gene trap contains the elements described above but also a second gene which is a selectable marker. This is controlled by a constitutive promoter (P) and has a downstream splice donor (SD) making polyadenylation dependent on the interrupted gene. Under selection, cells or animals will survive only if the construct has interrupted an endogenous gene, but it does not matter if the gene is not expressed.

short interfering fragments that induce the rapid degradation of any mRNA with the same sequence (see p. 192). This method has been used to inactivate nearly 95% of the genes in the nematode, which as stated above contains relatives of 60% of the known human disease genes and provides useful models for a number of disorders. RNA interference can also be used directly in human cells to study cell-level phenotypes. Many disease affecting basic cellular processes (e.g. xeroderma pigmentosum is a DNA repair deficiency) can now be modeled in cell culture by RNA interference. This represents perhaps the only current method that allows the large scale and direct analysis of the functions of human genes.

Further reading

POGM: Chapter 5 introduces high-capacity sequencing vectors. Chapter 7 provides an overview of sequencing technology and an introduction to sequence analysis and bioinformatics. Chapter 13 discusses the use of gene traps and insertional vectors in gene cloning and identification.

POGA: Chapters 3 and 4 discuss strategies for creating a physical clone map of the genome. Chapter 5 discusses the development of sequencing technology for genome applications, including the need for automation. Chapter 6 discusses bioinformatics and techniques for genome annotation. Chapters 7–11 provide more detail on the technology and applications of functional genomics including comparative genomics (Chapter 7), sequence and structural comparisons (Chapter 8), expression profiling in transcriptomics and proteomics (Chapter 9), the development of comprehensive mutant libraries (Chapter 10), and the analysis of protein interactions (Chapter 11).

Aebersold R, Cravatt BF (2002) Proteomics – advances, applications and the challenges that remain. *Trends Biotechnol* **20** (Suppl 12), S1–2.

Coelho PSR, Kumar A, Snyder M (2000) Genome-wide mutant collections: toolboxes for functional genomics. *Curr Opin Microbiol* **3**, 309–315.

Drewes G, Bouwmeester T (2003) Global approaches to protein–protein interactions. *Curr Opin Cell Biol* **15**, 1–7.

Lee KH (2001) Proteomics: a technology-driven and technology-limited discovery science. *Trends Biotechnol* **19**, 217–222.

Pandey A, Mann M (2000) Proteomics to study genes and genomes. *Nature* **405**, 837–846.

Stanford WL, Cohn JB, Cordes SP (2001) Gene-trap mutagenesis: past, present and beyond. *Nature Rev Genet* **2**, 756–768.

Zhu H, Snyder M (2003) Protein chip technology. *Curr Opin Chem Biol* **7**, 55–63.

A collection of recent reviews covering the technology and applications of proteomics and mutational genomics.

Celis JE, Kruhoffer M, Gromova I *et al.* (2000) Gene expression profiling: monitoring transcription and translation products using DNA microarrays and proteomics. *FEBS Lett* **480**, 2–16.

A thorough review covering the methodology of transcriptomics and proteomics and providing many examples of how these techniques are being applied in medical research and drug discovery.

Clark MS (1999) Comparative genomics: the key to understanding the Human Genome Project. *Bioessays* **21**, 121–130.

A paper demonstrating how comparative genomics is helpful in gene prediction and functional annotation.

Collins FS, McKusick VA (2001) Implications of the human genome project for medical science. *J Am Med Assoc* **285**, 540–544.

Subramanian G, Adams MD, Venter JC, Broder S (2001) Implications of the human genome for understanding human biology and medicine. *J Am Med Assoc* **286**, 2296–2307.

Van Ommen GJ (2002) The Human Genome Project and the future of diagnostics, treatment and prevention. *J Inherit Metab Dis* **25**, 183–188.

Three excellent reviews illustrating the expected medical benefits of the Human Genome Project.

Davies K (2001) *Cracking the Genome: Inside the Race to Unlock Human DNA*. Free Press, New York.

Green ED (2001) Strategies for the systematic sequencing of complex genomes. *Nature Rev Genet* **2**, 573–583.

Excellent coverage of genomic sequencing technology.

International Human Genome Sequencing Consortium (2001) Initial sequencing and analysis of the human genome. *Nature* **409**, 860–921.

McCain L (2002) Informing technology policy decisions: the US Human Genome Project's ethical, legal, and social implications programs as a critical case. *Technol Soc* **24**, 111–132.

A well-written and critical article which provides a history of ELSI programs and appraises their ability to fulfill the mandate set out during the Human Genome Project.

Stein L (2001) Genome annotation: from sequence to biology. *Nature Rev Genet* **2**, 493–503.

Comprehensive account of how to find genes in genomic DNA.

Various authors (1999) The Chipping Forecast. *Nature Genet* **21** (Suppl), 1–60.

Various authors (2002) The Chipping Forecast II. *Nature Genet* **32** (Suppl), 465–551.

Velculescu VE, Volgelstein J (2000) Analysing uncharted transcriptomes with SAGE. *Trends Genet* **16**, 423–425.

A collection of reviews charting the development and use of microarrays plus a short article on the use of sequence sampling.

Venter JC, Adams MD, Myers EW *et al.* (2001) The sequence of the human genome. *Science* **291**, 1304–1351.

A recent book providing a good overview of the Human Genome Project, and the two rival papers published by the publicly funded consortium and Celera Genomics.

CHAPTER THREE

Genomics and the challenge of infectious disease

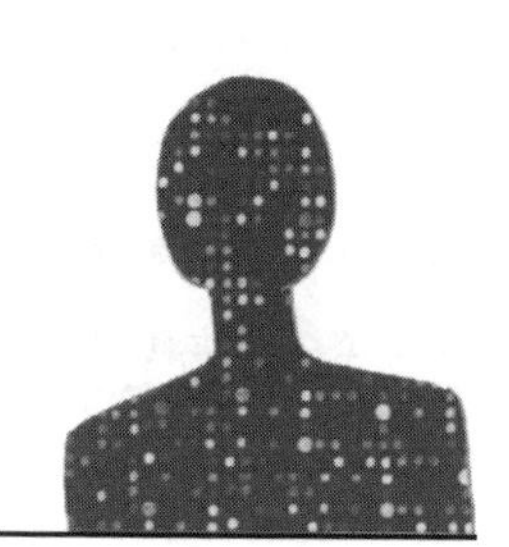

Microorganisms causing disease

Microbes that cause disease are often described as pathogenic or virulent. This suggests that they incite infections regardless of their quantity, their portal of entry, or the presence of other microorganisms; that is, pathogens are distinguished from nonpathogens by their expression of virulence factors. Although this concept appears to apply to certain microbes that cause disease in normal hosts, there are many others (opportunistic pathogens) for which it does not. Although virulence is a microbial characteristic, it can only be expressed in a susceptible host. Thus, so-called virulent organisms may be avirulent in a host with specific immunity, whereas a microbe that usually is avirulent can cause disease in immunocompromised hosts. From the perspective of the human host, the type and amount of damage caused is the only relevant outcome of the microbe–host interaction.

The microbes that cause disease fall into four categories: viruses, bacteria, fungi, and protozoa. Viruses differ from the other three groups of organisms in a number of key respects. First, they contain only one kind of nucleic acid (RNA or DNA) whereas cellular organisms contain both types. Second, they can be reproduced solely from their nucleic acid, i.e. viral nucleic acid is infectious. By contrast, cellular organisms are reproduced from the integrated sum of their components. Third, the different structural components of a virus are synthesized separately and then assembled into mature virus particles. Growth of cells occurs by an increase in the amounts of all their components during which the integrity of the cell is maintained. Fourth, viruses do not have ribosomes and are dependent on host cells for the provision of the machinery for synthesis of viral proteins. Finally, viruses do not encode any of the biochemical mechanisms for ATP synthesis.

It should be clear from the above that viruses are obligate intracellular parasites that are heavily dependent on host biochemistry for their replication. This feature of viruses has a number of consequences for human health care. First, all the viruses known to infect humans cause disease because of their effects on the host cell. Second, the production of antiviral drugs has proved difficult because most agents

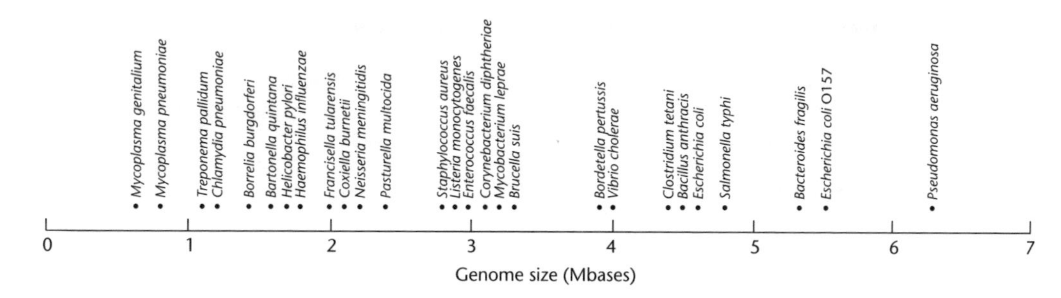

Fig. 3.1 The range of genome sizes found in different bacterial pathogens.

that prevent virus multiplication also interfere with essential cellular processes. For this reason, prevention of virus infections through the use of vaccines has been the most useful medical strategy. However, advances in our understanding of the molecular biology of viral nucleic acid replication are now permitting the development of clinically useful antiviral drugs (see p. 86).

A wide range of bacteria is capable of causing disease in man and animals. These pathogens include bacteria with some of the smallest known cellular genomes (mycoplasmas) and some of the largest bacterial genomes (*Pseudomonas aeruginosa*) (Fig. 3.1). Some of these bacteria are obligate intracellular parasites and genome sequencing has revealed the reason for this: the absence of genes for some of the central metabolic pathways (Table 3.1). Physiologic studies and genome sequencing also have revealed that although metabolically diverse bacteria can cause disease, they use common mechanisms to overcome host defenses (see p. 70).

Classically, two approaches have been used to fight bacterial infections: vaccination to **prevent** infections and antibiotics to **treat** infections. Vaccines are available for preventing infections by a range of bacteria (Table 3.2). However, with the exception of the tuberculosis (TB) and anthrax vaccines, all of them use killed bacteria and hence immunity is not long-lived. Molecular genetics and genomics are providing a range of methods for generating safe, attenuated vaccines and these are described in more detail later (see p. 76).

Originally, the term "antibiotic" was used to describe any microbial product that, in low concentration, could inhibit or kill other microorganisms. The archetypal example of a classic antibiotic is penicillin and many such antibiotics in use today were found by screening large numbers of microbial isolates for their ability to secrete substances that would inhibit the growth of one or more pathogens. These

Table 3.1 Some examples of missing metabolic reactions leading to obligate intracellular growth.

Organism	Missing biochemical capabilities
Mycoplasma genitalium	TCA cycle; biosynthesis of purines, pyrimidines, and amino acids
Treponema pallidum	TCA cycle; biosynthesis of purines, pyrimidines, and amino acids
Rickettsia prowazekii	Glycolysis; biosynthesis of purines, pyrimidines, and amino acids
Chlamydia trachomatis	TCA cycle; biosynthesis of purines and pyrimidines
Mycobacterium leprae	Partial loss of glycolysis and TCA cycle

Table 3.2 Bacterial pathogens for which a vaccine is available.

Organism	Disease
Bacillus anthracis	Anthrax
Bordetella pertussis	Whooping cough
Clostridium tetani	Tetanus
Corynebacterium diphtheriae	Diphtheria
Haemophilus influenzae	Meningitis
Mycobacterium tuberculosis	Tuberculosis
Neisseria meningiditis (some strains)	Meningitis
Salmonella paratyphi A & B	Paratyphoid A & B
Salmonella typhi	Typhoid
Streptococcus pneumoniae	Pneumonia
Vibrio cholera	Cholera

antibiotics can be used clinically because the biochemical processes that they target are absent from mammalian cells. For example, penicillin and a number of other antibiotics interfere with bacterial cell wall synthesis whilst others target components unique to prokaryotic ribosomes. Today, the term antibiotic is used to describe any natural, semisynthetic or wholly synthetic antimicrobial compound that is used therapeutically and the process of identification of candidate molecules occurs in exactly the same way as any other drug. Gene manipulation techniques and genomics methodologies are facilitating our understanding of the mechanisms of bacterial pathogenicity and the origins of epidemics of drug resistance. These advances are helping in the selection of new biochemical targets for antibiotics and are described later in this chapter. The way in which new antibiotic candidates are developed as drugs is discussed in Chapter 7.

Whereas many different bacteria can cause infections in humans and other mammals, the range of clinically significant fungi is much more restricted. Most people are familiar with superficial mycoses such as athletes' foot and ringworm. Although these skin infections are esthetically displeasing the clinical consequences are minor. By contrast, systemic mycoses are life-threatening infections. Relatively few fungal species are primary pathogens, i.e. invade and colonize healthy individuals. The principal ones are *Histoplasma capsulatum*, *Coccidioides immitis*, *Paracoccidioides brasiliensis*, *Blastomyces dermatitidis*, and those fungi that infect skin (the dermatophytes). Other major fungal pathogens, such as *Candida* sp., *Aspergillus fumigatus*, *Pneumocystis carinii*, *Rhizopus oryzae*, and *Fusarium solani*, are opportunistic pathogens. These fungi cause life-threatening infections in patients with severely compromised immune systems. There has been a big increase in the numbers of such immunocompromised patients due to increased cancer chemotherapy and the numbers of cases of AIDS. Opportunistic pathogens illustrate clearly the point made at the beginning of this chapter, i.e. infection is the outcome of a microbe encountering a susceptible host. *Cryptococcus neoformans* can cause community acquired infections but mostly affects immunocompromised patients. For most of these pathogens the commonest route of infection is inhalation of spores.

Table 3.3 The common protozoal infections.

Organism	Disease	Transmission	Distribution
Cryptosporidium	Severe diarrhea	Contaminated water	Worldwide*
Giardia	Severe diarrhea	Contaminated water	Worldwide*
Leishmania	Leishmaniasis	Biting insects	Hot/tropical countries
Plasmodium sp.	Malaria	Anopheline mosquitoes	Hot/tropical countries
Pneumocystis carini	Pneumonia	Inhalation	Immunocompromised patients
Toxoplasma	Toxoplasmosis	Various	Worldwide. Growing problem in immunocompromised patients (e.g. AIDS)
Trypanasoma brucei	Sleeping sickness	Tsetse fly	Africa
Trypanasoma cruzi	Chagas disease	Oral ingestion of insect feces	South America

* Infections that are rare in developed countries but common in underdeveloped countries.

Finding antibiotics that treat fungal infections is harder than finding antibacterials because both the target and the host are eukaryotic and there are fewer biochemical differences between them than between bacteria and mammals. An obvious antifungal target is the cell wall but cell wall inhibitors have proved to be of limited clinical utility. To date, the most successful antifungal compounds have been the azoles which target steps in fatty acid biosynthesis that are absent from higher plants and animals. As with the search for new antibacterials, genomics should increase the number of potential targets for new antifungal compounds.

Protozoa differ from the other classes of pathogen in that their significance is largely confined to tropical and/or underdeveloped countries (Table 3.3). This in turn is a reflection on their mode of transmission. Most protozoa are spread either by biting insects (mosquitoes, sand flies) or by ingesting contaminated water, although a few are transmitted sexually (e.g. *Trichomonas vaginalis*). Because protozoal diseases are relatively insignificant in western countries, very little effort has been extended in developing effective drugs to treat them. Because of this disinterest, malaria kills more people in the world than any other microorganism. For many years the most effective remedies for preventing and treating diseases such as malaria were purified preparations derived from traditional native remedies, e.g. quinine from the bark of the *Cinchona* tree. Even today, most of the antiprotozoal drugs are synthetic derivatives of the active ingredients from these plant-derived medicines. Fortunately, genomics is coming to the rescue and novel ways of fighting malaria are beginning to emerge (see p. 83).

Where do new diseases come from?

From time to time new infectious disease agents are discovered. For example, a large number of people attending an American Legion convention in Philadelphia in 1976 developed pneumonia and a significant proportion of them died. Microbiologic examination of lung tissue taken at autopsy resulted in the isolation of a new bacterium subsequently named *Legionella pneumophila*. Why had this organism not been isolated before? The reason is simple. *Legionella* are ubiquitous in natural freshwater environments although they normally constitute only a small proportion of the total bacterial load. However, unlike most bacteria, they can grow and survive at elevated temperatures and hence they thrive in industrial and domestic water systems. Many of these water systems produce aerosols, e.g. showers, whirlpool spas, and air conditioning systems, and the infected droplets readily penetrate the alveolar airspaces of the lung. In essence, changes in lifestyle created conditions that favored the spread of *Legionella* in a form that facilitates infection of susceptible hosts. Had it not been for the numbers of individuals infected simultaneously, the identification of this "new" pathogen might not have occurred.

Campylobacter species are a major cause of bacterial food poisoning. Like *Legionella*, their involvement in human disease was not fully appreciated until the late 1970s. There were a number of reasons for this. First, there were a rapidly growing number of cases of food-associated enteritis that were not due to *Salmonella* but for which no causative agent had been isolated. Many of these cases of enteritis were linked with the consumption of chicken and *bacterial* infection was suspected. Second, *Campylobacter* are very fastidious organisms and their laboratory cultivation requires specialist media. Only when the number of cases of enteritis of unknown etiology reached a high level was significant attention paid to identifying the causative agent. It is worth noting that the increase in enteritis paralleled the growth in the consumption of chicken. The growing demand for chicken in turn had fuelled the growth in high-density (battery) farming methods, a practice that greatly facilitates the spread of *Campylobacter*. Once again, changes in lifestyle facilitated a change in patterns of infectious disease.

Most of the emerging viral diseases are not new diseases in global terms, they are just new to the western world. These diseases always have been endemic in third world countries and probably afflicted visitors from developed countries. However, a number of social, economic and lifestyle factors has led to their increase. First, the mass destruction of tropical forests has encouraged canopy-dwelling animals to adapt to life on the ground. Many of these animals are reservoirs of human diseases and when close to the ground facilitate insect transmission. Second, rapid transport means that infected westerners can return to their home base for treatment whereas previously they would have died of "fever" whilst "in the field." An added factor here is the growing trend to take holidays in exotic locations. Third, in many developing countries there has been a large scale movement from rural communities to cities and this has increased the awareness of health authorities of the extent of these infections. Finally, global warming means that many insect vectors are spreading to more developed regions where they previously were unknown. Nevertheless, there are several viral diseases whose origin is unclear. For example, Ebola is highly infectious and nearly always fatal. It appears to be a recent disease but why this should be is not known. There must be a reservoir of infection but all attempts to find it have failed. Similarly, there is an epidemic of AIDS due to HIV infection. Although the factors promoting its spread are known (sexual practices), the origins of the disease are still contentions.

Identifying the causative agent of a disease

The first and foremost reason for identifying the microbial agent responsible for disease in any particular patient is to select the most appropriate therapy. The second reason is to collect information for public health purposes. For example, a persistent low level of food poisoning in any community is to be expected, but a sudden increase in the numbers of illness due to *Salmonella typhimurium* or hepatitis A virus suggests an outbreak of food poisoning from a particular source that needs to be traced.

The primary identification of a microbial disease is undertaken by a physician, often a general practitioner, on the basis of symptoms exhibited by the patient. Depending on the nature of the infection, the primary diagnosis may be confirmed by laboratory examination of patient specimens. The accuracy of diagnoses made on the basis of symptoms is high for those diseases that are common in any particular community or for which the doctor has specialist knowledge. It is "exotic" diseases that the physician has never encountered before that pose the real difficulty. These could be diseases caused by microbes only found in tropical regions or underdeveloped countries or could be ones that once were common but have largely been eliminated in the western world. For example, many western GPs have never seen a case of tetanus but a doctor in the Indian subcontinent would easily recognize the symptoms. Similarly, a physician from West Africa would recognize the symptoms of malaria in a feverish patient but they might confuse a GP in Northern Europe. Acts of bioterrorism also could result in epidemics of "exotic" diseases (Box 3.1). In the absence of bioterrorism, "exotic" infections are most likely to arise from travel to underdeveloped regions. By determining the recent travel history of the patient and the symptoms exhibited it should be possible to narrow down the list of potential infecting agents (Fig. 3.2).

Many different kinds of samples can be presented to a clinical microbiology laboratory including blood, urine, feces, sputum, tissue, etc., and the usual requirement is the identification of any infectious agents present and, if appropriate, their antibiotic sensitivity. If the identification simply is to confirm a clinical diagnosis then a fairly limited number of tests will be done on the sample and these are unlikely to involve molecular methods. Where there has been no presumptive identification of the infectious agent then a much more extensive set of tests will be undertaken (Fig. 3.3), especially if the patient is seriously ill. A similar test procedure will be used if a highly infectious agent is suspected, e.g. smallpox, plague, anthrax, but the testing will be undertaken in a specialist, high-containment facility.

The detection in a serum sample of an antibody to a pathogen is a good indication of recent exposure to the organism in question, especially if the antibody titer is rising. There are two basic methods of detecting antibodies. In the first of these, one detects the formation of immunoprecipitates when the test sample is mixed with the appropriate antigen. In the second, the antigen is immobilized on a solid surface and then flooded with the test serum. After washing off any unbound protein the presence of antibody bound to the antigen is detected by the addition of fluorescently labeled anti-human immunoglobulin antibody.

Box 3.1 Biological warfare agents

There are many recorded cases in the literature of the use of microbes as biologic weapons. In recent years the primary concern has been the mass production of biologic warfare agents by certain "rogue" nations and the formulation of these agents in a format suitable for delivery in missiles ("weaponization"). However, the major threat so far has come from local terrorists. For example, in 1993 the Aum Shinrikyo sect in Japan released anthrax spores. More recently, in October 2001, five people died and another 12 were infected when anthrax spores were sent to them in the US mail. Then, in January 2003, police in London found a "factory" isolating ricin, a highly potent toxin, from castor oil seeds. These events have highlighted the need for countries to have a strategic plan for responding to bioterrorism.

The US government has divided potential biowarfare agents into three categories based on perceived risk (Table B3.1).

Because infections with the category A and B organisms listed above are so rare in the industrialized world, most primary care physicians would not recognize the symptoms. In the US anthrax cases, diagnosis was greatly facilitated by the fact that victims were exposed to a "white powder" suggesting that a chemical or biologic agent might have been involved. Even when the disease symptoms are correctly diagnosed there is the additional problem that there are no standard methods for determining the antibiotic sensitivities of any isolates. Although the most suitable antibiotics for use with wild-type strains (natural isolates) are known, defectors from the bioweapons program in the former Soviet Union have revealed that they developed strains resistant to all the usual antibiotics.

Table B3.1

Category	Characteristics	Examples
A	Can be easily disseminated or transmitted from person to person Can cause high mortality and has the potential for major public health impact Might cause public panic and social disruption Requires special action for public health preparedness	*Bacillus anthracis* (anthrax) *Yersinia pestis* (plague) *Francisella tularensis* (tularaemia) Variola major (smallpox) Viral hemorrhagic fevers (e.g. Ebola, Marburg)
B	Are moderately easy to disseminate Will cause moderate morbidity and low mortality Require enhancement at national level of diagnostic and surveillance capabilities	*Brucella* sp. *Burkholderia mallei* *Burkholderia pseudomallei* Viral encephalitis
C	Emerging pathogens with major public health impact that could be engineered for mass dissemination because of availability and ease of production	Hantahviruses Nipahvirus

The classic methods of detecting antigens are culture of the infectious agent, microscopy of wet and stained specimens, and the detection of pathogen-specific antigens by ELISA (enzyme-linked immunosorbent assay) or other immunologic techniques. Molecular techniques such as the PCR are not used routinely unless the infection is life threatening or highly infectious. There are two reasons for this. First, the PCR is so sensitive that it is easy to amplify very small amounts of

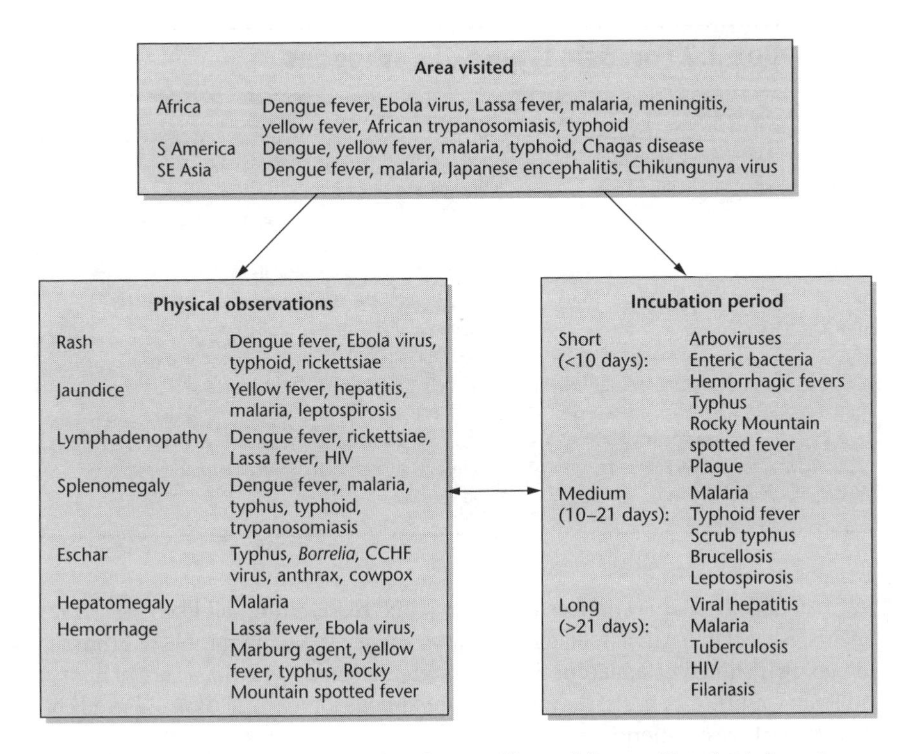

Fig. 3.2 Diagnostic criteria used in the identification of "exotic" diseases. Note that information on the precise countries visited is far more important than the general geographic area.

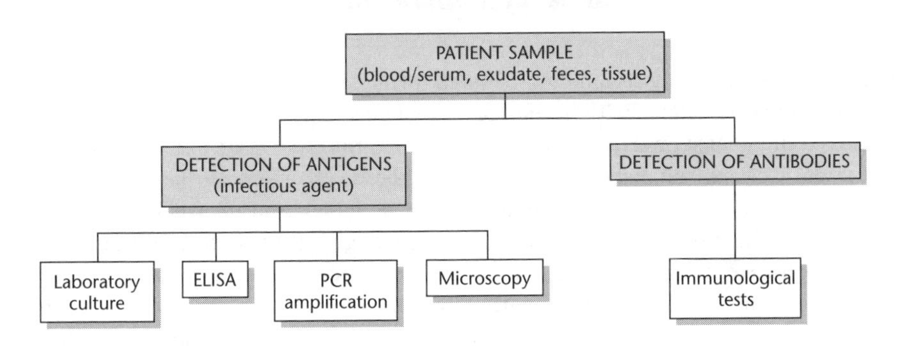

Fig. 3.3 Types of laboratory tests done to identify "exotic" diseases.

contaminating DNA. Therefore great care must be exercised in all laboratories using the technique routinely otherwise wrong diagnoses will be made. Second, the PCR is ideally suited to yes/no decisions, i.e. does this sample contain *Bacillus anthracis* (anthrax) DNA or does it not. It is much less suited to determining which organism out of many alternatives is present in a sample. The reason for this is that the PCR works best when only one or a few pairs of primers are used in the

Box 3.2 Forensic typing of pathogens

Most people are familiar with the now widespread use by forensic scientists of DNA fingerprinting for the identification of rapists and other criminals. In a remarkably similar way, the analysis of DNA polymorphisms in microorganisms (microbial forensics) is being used to infer the origin, relationships or transmission route of a particular isolate, particularly if it is associated with malicious activities.

One of the earliest uses of microbial forensics was the investigation of the allegation that a number of patients in Florida contacted HIV from their dentist. Proof that this had indeed occurred was provided by comparison of the sequences of amplified viral fragments from the patients and the dentist. Another noteworthy application was the investigation of the polymorphisms in the genome of the *Bacillus anthracis* spores released in Japan in 1993 by the Aum Shinrikyo cult. Analysis of multiple-locus variable number tandem repeats showed that the strain used in this localized bioterrorist attack was the Sterne 34F2 strain that is used in Japan for animal prophylaxis against anthrax. By contrast, the molecular investigation of the West Nile virus outbreak in the USA in 1999 identified only a single strain from birds and humans in New York. This strain had greatest similarity to an isolate from a dead goose in Israel and it was concluded that the virus was of natural origin.

test reaction. If a large number of primer pairs, representing all likely infectious agents, is used in a single reaction then interference is a major problem. From the foregoing it should be apparent that although molecular techniques are of limited utility in routine diagnosis, they could be invaluable for forensic (Box 3.2) and epidemiologic investigations.

Molecular epidemiology

Microbial epidemiologists monitor the spread of microorganisms, principally bacteria and viruses, associated with human or animal infectious diseases at levels ranging from a single host or ecosystem (e.g. hospital) to the worldwide environment. On the basis of epidemiologic investigation, public health risks can be determined, interventions in the spread of disease can be designed, and their efficacy can be assessed. The fundamental principle of epidemiology is that all the strains isolated from a disease outbreak should share a certain set of characteristics that differ from those of isolates or strains not involved in the outbreak. These characteristics are determined by typing and may be phenotypic or genotypic. Classically, bacterial strains were characterized by phenotypic methods using markers such as the ability to ferment different sugars, the presence or absence of particular enzymes, sensitivity to a range of bacteriophages, and polyacrylamide gel electrophoresis (PAGE) of protein extracts. Today, there is a much greater emphasis on methods that identify DNA polymorphisms in the different isolates.

Many different molecular methods have been developed for typing microorganisms but the ones in widespread useage are RAPD (random amplification of polymorphic DNA), AFLP (amplified fragment length polymorphisms), PFGE (pulsed field gel electrophoresis), ribotyping, and multilocus sequencing. The first three of

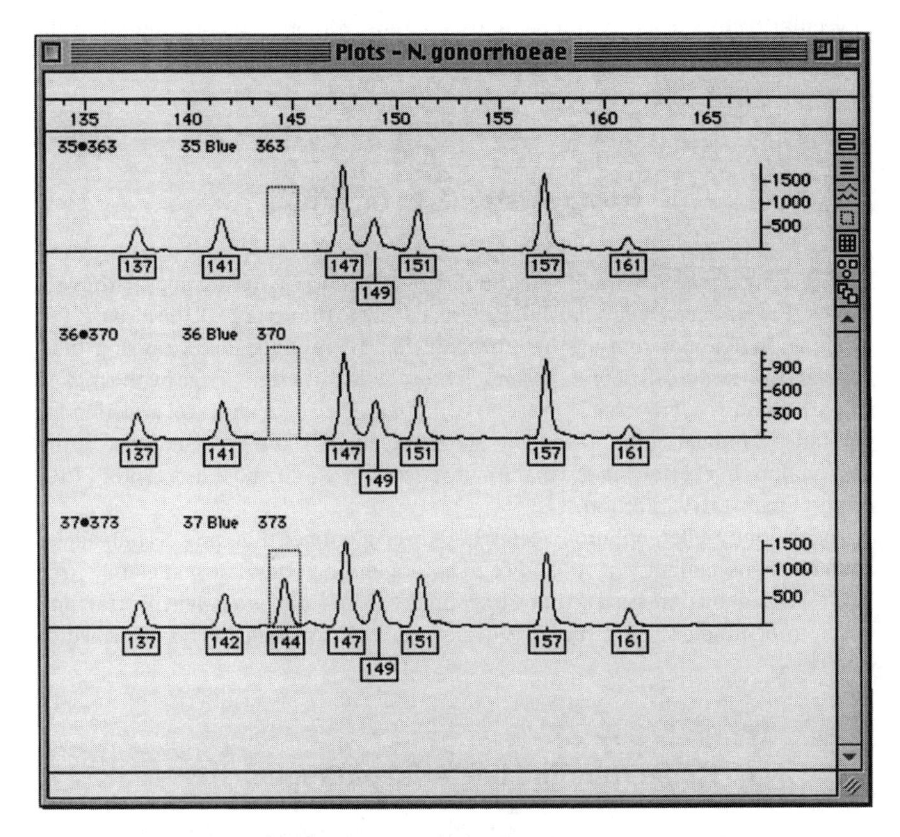

Fig. 3.4 Fluorescent AFLP profiles generated from three *Neisseria gonorrhoeae* isolates. A 30 base pair window (135–165 bp) is shown. The boxed numbers below the fluorescent peaks are the sizes in base pairs. The scale to the right of each profile indicates fluorescent units. The top two profiles were generated from isolates of epidemiologically linked cases and the lower profile is a serologically identical but unlinked case. The peak at 144 bp highlighted in the lowest profile separates the isolate from the two identical profiles above.

these methods identifies polymorphisms in the whole genome whereas the last two focus on a small group of genes. RAPDs and AFLPs both involve cleavage of DNA with restriction endonuclease, amplification of selected fragments by the PCR, and the resolution of the fragments by gel electrophoresis. Identical isolates should produce the same pattern and one that is different from unrelated isolates of the same organism (Fig. 3.4). The detailed methodologies are quite different and suffice it to say that the AFLP method gives much more reproducible patterns than the RAPD method. In the PFGE method, genomic DNA is cleaved with a restriction endonuclease and the fragments resolved by pulsed field gel electrophoresis. In ribotyping, a detailed genomic analysis is made of the 16S and 23S rRNA genes and the spacer region between them. In multilocus sequence typing, internal fragments from a number of housekeeping genes are amplified and sequenced and the sequences from the different isolates compared. Because of the discriminatory power of these molecular techniques they are being used forensically in cases of criminal negligence and to trace the source of disease outbreaks that might be linked to bioterrorist activities (Box 3.2).

Host resistance to infection

Not all individuals are equally susceptible to infection by a microbial pathogen. Rather, the host genotype contributes significantly to the severity of the disease. For example, individuals that are heterozygous for the mutant alleles causing thallassemia and sickle cell anemia are much more resistant to challenge by the malarial parasite. It is even possible that the two diseases arose by natural selection for resistance to malaria. Another example of the effect of the host genotype is the observation that certain alleles of the gene encoding the chemokine receptor CCR5 protect against HIV infection.

Today, many different human genes have been identified that have certain alleles conferring susceptibility or resistance to pathogens, e.g. the HLA gene family (see p. 106). What has emerged is that susceptibility to infection is a polygenic trait and needs to be studied using specialized methodologies such as those described in Chapter 4.

Understanding bacterial pathogenicity

Bacteria use a variety of biochemical mechanisms to gain access to their niche on or within a host, to colonize that niche, to escape the host defences, and to exclude competing organisms. Many of these virulence factors and their regulatory elements can be divided into a smaller number of groups based on the conservation of similar mechanisms. These common themes in bacterial virulence are not restricted to pathogens of man and other warm-blooded animals but also are found in plant pathogens and bacteria such as *Rhizobium* sp. that have a symbiotic relationship with plants.

With the exception of those few bacteria that synthesize toxins in the absence of eukaryotic cells, e.g. *Clostridium botulinum*, contact between a bacterium and its site of infection is an essential prerequisite for pathogenicity. Bacteria possess specialized factors required for attachment to the host and these are either type IV pili (fimbriae) or nonpilus adhesions. Only a few basic designs of these molecules have been observed. There is considerable sequence similarity between the nonpilus adhesions of *Hemophilus* and *Moraxella* (human pathogens) and those of the plant pathogen *Xylella*. Similarly, the type IV pili of *Pseudomonas aeruginosa*, *Neisseria* sp., *Vibrio cholerae*, *Moraxella* sp. and enteropathogenic *E.coli* have common structural and genetic features. For example, the pilin subunits have a conserved, unusual amino-terminal sequence that lacks a classic leader sequence and that is removed by a specific leader peptidase. Several possess methylated amino termini on their pilin molecules and usually contain pairs of cysteines that are involved in intra-chain, disulfide bond formation near their carboxy termini.

Some bacteria possess the means to enter host cells and replicate there and this requires special determinants called invasion factors. These factors frequently subvert the signaling systems of the host thereby promoting internalization of the bacterium by cells not normally proficient for phagocytosis, e.g. epithelial cells. Two general mechanisms of internalization have been discovered and both involve structural mimicry. One of these mechanisms involves the production by the pathogen of homologs of host proteins. For example, *Yersinia* sp. and *Salmonella* sp. produce tyrosine phosphatases with considerable sequence and structural homology to eukaryotic phosphatases. Because tyrosine phosphorylation does not commonly occur in bacteria it is probable that these molecules have evolved specifically to modulate host cellular functions. The second mechanism involves effectors that have no obvious amino acid sequence similarity to host cell proteins but which have similar three-dimensional structures. This kind of mimicry is very common in the enteric pathogens as well as *Listeria*, *Helicobacter*, and *Pseudomonas* sp.

Part of the pathogenic process involves avoiding destruction by the host defenses. In some cases this is achieved by adopting an intracellular lifestyle whereas in others the bacterium synthesizes extracellular polysaccharides that are poor immunogens and have antiphagocytic properties. If bacteria cannot avoid attack by immunoglobulin molecules then they may synthesize specific enzymes to destroy them. Should phagocytes engulf the pathogen it may be able to synthesize protective factors to overcome the conditions of oxidative stress, low pH, and attack by antimicrobial defense peptides.

Toxins are the best known of the bacterial virulence factors. In some cases these are produced to enable the pathogen to degrade the membrane of the phagocytic vacuole allowing it to escape into the host cytoplasm. In other cases, a noninvasive pathogen may use its toxins to effect the release of nutrients from the host. Again, there is a commonality of mechanisms. In Gram-negative plant and animal pathogens, these toxins are delivered into host cells via the type IV pili described earlier or by a complex protein secretion system termed type III. Among the 20 or so proteins involved in type III secretion, at least nine are conserved between different bacterial species and probably constitute the core components of the secretion apparatus. Similarly, there are only a few types of toxins, despite a large number of different host targets. RTX (repeats in toxins) are produced by a wide variety of Gram-negative bacteria and fall into two categories: the hemolysins that affect a variety of cell types and the leukotoxins that are more cell- and species-specific. The characteristics of these toxins and their corresponding genes are summarized in Table 3.4. Other virulence factors that are easily detected during whole genome sequencing projects are proteins involved in iron sequestration and antibiotic synthesis. Iron is essential for virulence but is never free in the cytoplasm or the extracellular matrix. In the case of antibiotics, their function is believed to be the suppression of other bacteria that could compete for nutrients released as a result of host cell destruction.

Comparative genetics can be used to recognize virulence determinants in a newly sequenced organism. Genes with significant homology to well-characterized

Table 3.4 Generic properties of RTX toxins.

Feature	Properties
Operon structure	Four genes designated *rtxC*, *rtxA*, *rtxB* and *rtxD* in that transcriptional order
Protein structure	The toxin is the product of the *rtxA* gene and contains tandemly repeated nonapeptides with the consensus sequence GGXGXDX[L/I/V/W/Y/F]X where X is any amino acid. This nonapeptide is repeated 6–40 times
Post-translational modification	RTX toxins require post-translational modification to become biologically active and this occurs by acylation of the *rtxA* gene product by the *rtxC* gene product
Secretion	RTX toxins do not have canonical signal sequences but are secreted with the help of the *rtxB* and *D* gene products

virulence genes in other organisms are likely to be involved in virulence as well. However, caution must be exercised if the homology is to a "putative" virulence gene because many of these subsequently are found to have no role in infection. Another indicator is if a gene product is present in virulent strains but absent from nonvirulent or less virulent isolates of the same species. Evidence for a role in virulence also is suggested if a gene product has numerous amino acid substitutions in different isolates.

Pathogenicity islands

The fact that unrelated pathogens use very similar mechanisms for invading the host and that there is considerable protein sequence homology for each type of pathogenicity determinant suggests that there has been extensive horizontal transfer of virulence genes. Support for this comes from the identification of pathogenicity islands (PAIs) in Gram-positive and Gram-negative bacteria whose genomes have been completely sequenced. PAIs are large chromosomal regions (35–200 kb) encoding several virulence gene clusters that are present in all pathogenic isolates and generally absent from nonpathogenic isolates. Often they can be recognized in genome sequences by the fact that their percentage G + C content differs from that of the rest of the genome. Of particular interest is the observation that PAIs encode an integrase, are flanked by direct repeats, and are inserted into the bacterial chromosome adjacent to tRNA genes. In this respect, PAIs resemble temperate bacteriophages and could have been acquired by new hosts as a result of transduction or via conjugative transposons. In some cases, PAIs are located on plasmids and probably got there by excision from the chromosome and integration into an "avirulent" plasmid resident in the same cell. Examples of plasmids carrying PAIs have been found in *Shigella* and *Yersinia* species and *Bacillus anthracis*. The latter is a particularly interesting example since the main chromosome of *B. anthracis* is very similar to that of *B. cereus* and *B. thuringiensis*. The latter two organisms are not pathogens, although they can produce a toxin, and both lack the plasmid carrying the PAI.

Comparative genomics and genome plasticity

Until the advent of whole genome sequencing, closely related bacteria or even different strains of the same bacterium could only be compared by relatively crude methods. For example, biochemical characterization (metabolic profiles) and DNA fingerprinting in reality only examine a very small part of the total genome. For a complete analysis it is necessary to sequence the entire genomes, including plasmids, of the organisms under study. This is possible and has been done on a number of occasions but is very laborious. An alternative technique is to sequence one genome and use the information obtained to construct a DNA array that can be used to detect the presence or absence of genes in other related genomes. To compare strains, total DNA is prepared, labeled with a fluorophore, and hybridized to the array. Different genomes can be compared by labeling the DNA samples with different fluorophores. The DNA samples are mixed and after hybridization the intensities of the two fluorophores are compared for each locus represented on the microarray (see Chapter 2). The disadvantage of this technique is that it can only detect deletions in a strain relative to the reference strain whose DNA was used to construct the microarray.

A large number of comparative analyses have been done on either different strains of the same species or different species of the same bacterium (Box 3.3). What these analyses have shown is that there is extensive sequence conservation but that within a genus or species there can be 5–25% of DNA unique to each isolate. For example, two different clinical isolates of *Helicobacter pylori* each had 7% of unique DNA and different *Salmonella* serovars have 10–12% of unique DNA. In most instances this unique DNA is scattered throughout the genome and is not represented by large blocks of DNA (except for the presence of PAIs in virulent strains relative to avirulent strains). Comparative analyses of strains are particularly useful in studying the emergence of new diseases and new disease phenotypes (Box 3.4).

The presence of so much unique DNA in different strains prompts two questions. First, what is the function of the unique DNA and, second, how does this genome variability arise? Almost certainly, the DNA that is unique to particular isolates encodes genes that enable the isolates to more effectively colonize the niche that they occupy. For example, vancomycin-resistant enterococci (VREs) are widespread in the environment and hospitals but only those strains carrying a variant of the *esp* (enterococcal surface protein) gene are found in hospital outbreaks. It should be noted that, in some cases, virulent isolates **lack** regions of DNA that are present in avirulent strains. For example, enteroinvasive *E. coli* and *Shigella* strains lack a chromosomal region that encodes *cadA* (lysine decarboxylase gene) and *ompT*. The lack of these genes enhances pathogenicity because cadaverine, a product of *cadA*, inhibits the enterotoxin activity of these strains

Box 3.3 Genomic variation in tuberculosis vaccine strains

Bacille Calmette-Guérin or BCG is the only vaccine currently available for use in preventing tuberculosis. It was developed in the early 1900s by attenuation of a virulent strain of *Mycobacterium bovis* by serial passage. It was first administered in 1921 to a newborn child whose mother had died of TB and who was going to live with a grandmother suffering from the disease. This individual remained free of TB throughout his life. A further 969 children were vaccinated over the next 6 years and their mortality from TB was only 10% of that of a control group. As a result, in 1928 the League of Nations recommended widespread vaccination with BCG. Although BCG is widely used throughout the world today its efficacy varies widely.

One explanation for the variation in efficacy of the vaccine could be the development of genetic differences as a result of different storage and culture conditions. Ever since the 1940s, differences in cultural properties have been noted in the BCG isolates held by different vaccine manufacturers. Using genomic analysis it has been possible to identify at least seven genetic events that have led to the various vaccine substrains in use today. These genetic events include deletions, duplications, and the introduction of a single nucleotide polymorphism (Fig. B3.3).

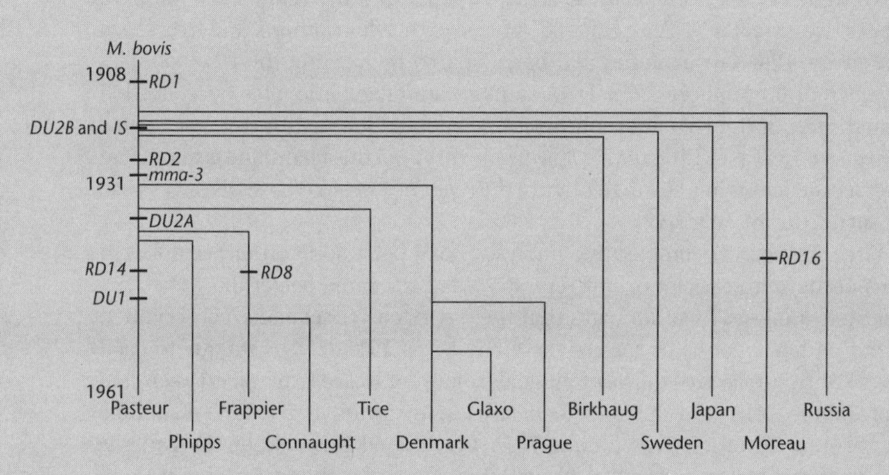

Fig. B3.3 The genealogy of different BCG strains. The vertical axis represents time and the horizontal axis shows the different locations where the BCG strain has been propagated. *RD1, RD2, RD8, RD14* and *RD16* are different deletions in the genome. *IS* represents the loss of the IS6110 sequence, and *DU1, DU2A* and *DU2B* represent three different genome duplications. A single base change designated *mma-3* interferes with the synthesis of methoxymycolic acids. (Adapted from Behr *et al.* 1999.)

and the *ompT*-encoded protease interferes with the VirG protein required for intercellular spread.

Analysis of sequenced genomes has shown that the main chromosome of most of them is heavily populated with temperate phages, transposons, insertion sequences, and small repetitive DNA sequences. In addition, many strains carry one or more plasmids. As a whole, these elements can promote genetic exchange and rearrangement via transformation, transduction, conjugation, and homologous and illegitimate recombination, thereby contributing to the observed genome plasticity.

Box 3.4 Insights into the origin of methicillin-resistant *Staphylococcus aureus* (MRSA) and the toxic shock syndrome epidemic

MRSA strains were first reported in 1961, soon after methicillin entered clinical use. Two very different hypotheses have been formulated to explain the origin of MRSA strains. The association of the *mecA* (penicillin binding protein) gene with diverse genetic lineages of *S. aureus*, together with data indicating that the *mec* gene was horizontally transferable in the laboratory, led to the hypothesis that MRSA had evolved many independent times. By contrast, data obtained from the study of MRSA by RFLP analysis with probes for the *mecA* gene were interpreted to mean that a susceptible strain of *S. aureus* acquired the *mecA* gene to become an MRSA and then diversified genetically. Comparative genomic analysis has shown that *S. aureus* strains are converted to MRSAs when they acquire a large (21–67 kb) genetic element known as staphylococcal cassette chromosome *mec* (SCC*mec*). This is a unique genetic element that contains no phage-related genes or transposases. Rather, it encodes two recombinases that facilitate integration and excision of the element from a precise chromosomal location. Three different types of SCC*mec* have been identified in MRSA strains and two others in *S. hemolyticus* and *S. hominis*. This strongly suggests that MRSA strains have arisen multiple independent times by lateral transfer of the *mecA* gene into susceptible precursors.

In the 1970s there was an epidemic of urogenital-associated toxic shock syndrome (TSS) in women associated with the introduction of a new kind of tampon. This prompted a debate: was the epidemic caused by a change in environment (the new tampon) or the rapid geographic dissemination of a new hypervirulent strain? Comparative genome analysis was undertaken on a selection of TSS strains of *S. aureus* isolated from different parts of the world and which exhibited the same multilocus enzyme electrophoresis pattern. The genomic analysis showed that although they all had shared a common ancestor they were not genetically identical and the last ancestor had not been very recent in evolutionary time. These data indicate that the epidemic was caused by a change in the host environment (new tampon) rather than rapid intercontinental spread of a new hypervirulent (TSS) strain.

In contrast to the above studies, comparative genome analysis of serotype M18 group A *Streptococcus* strains associated with acute rheumatic fever in the west-central USA over a period of 13 years showed that the strains were almost genetically identical. Given that the isolates contained numerous mobile genetic elements this observation suggests that they have an optimized genetic configuration for causing rheumatic fever in humans.

Combating infectious disease

The usual way of treating bacterial and fungal infections is with antibiotics. As is well known, certain microbes quickly develop resistance to the antibiotics in current use and this means that new antibiotics are required. Rather surprisingly, the recently introduced oxazolidinones were the first new class of antibiotic for over 25 years and already resistance is being found in clinical isolates. The traditional way of obtaining new antibiotics was the screening of natural isolates of microbes but the preferred way today is to identify new cellular targets and screen chemical libraries for inhibitory molecules. As will be seen later, genomics is facilitating the development of methods for identifying key genes involved in pathogenesis that could act as suitable targets for new antibiotics.

Whereas antibiotics are used to treat infections, vaccines are used to prevent infections in the first place. An effective vaccine against a bacterial pathogen will afford protection regardless of the antibiotic resistance status of the bacterium and this illustrates the benefits of prevention as opposed to cure. A vaccine works by generating humoral and/or cell-mediated immunity that prevents the development of disease upon exposure to the corresponding pathogen. This is accomplished by presenting pertinent antigenic determinants to the immune system in a fashion that mimics that in natural infections. Conventional viral vaccines consist of inactivated, virulent strains or live, attenuated strains but they are not without their problems. For example, many viruses have not been adapted to grow to high titer in tissue culture, e.g. hepatitis B virus. There is a danger of vaccine-related disease when using inactivated virus since replication-competent virus may remain in the inoculum. Finally, attenuated virus strains have the potential to revert to a virulent phenotype upon replication in the vaccinee. This occurs about once or twice in every million people who receive the polio vaccine. Genomics and gene manipulation techniques have been used in a variety of ways for the generation of vaccine candidates that do not have these problems.

Novel routes to vaccines

One alternative approach to the development of live vaccines is to start with the food poisoning organism *Salmonella typhimurium*. This organism can be attenuated by the introduction of lesions in the *aro* genes that encode the biosynthesis of aromatic amino acids. Whereas doses of 10^4 wild-type *S. typhimurium* reproducibly kill mice, *aro* mutants do not do so even when applied at doses a million-fold higher. However, the mutant strains can establish self-limiting infections in mice and can be detected in low numbers in organs such as the liver and spleen. Such attenuated strains of *S. typhimurium* can act as carriers of heterologous antigens because they can be delivered orally and because they can stimulate humoral, secretory and cellular immune responses. An alternative attenuated host that has been investigated is the Bacille Calmette-Guerin (BCG) vaccine strain of *Mycobacterium bovis*.

A different procedure for attenuating a bacterial pathogen to be used as a vaccine has been demonstrated in *Vibrio cholera*. The enterotoxin produced by this organism has two subunits, A and B. Using recombinant DNA technology, the A1 domain of the A subunit was deleted. The resulting strain produces the immunogenic B subunit of cholera toxin but is unable to produce the essential A subunit and has been shown to be safe and immunogenic in clinical trials. Undoubtedly, whole genome sequencing will identify other genes that can be deleted in pathogens to generate nonrevertible attenuated strains.

Genome sequencing has stimulated yet another approach to vaccine development as exemplified by the group B meningococci. These bacteria are notorious for displaying extensive variation in surface-exposed proteins, making it difficult to select vaccine candidates. Analysis of the meningococcal genome sequence identified over 500 proteins as potential protective antigens. Genes for 350 of these antigens were cloned in *E. coli*, the antigens expressed and purified, and then used

Table 3.5 Some heterologous antigens expressed by vaccinia virus recombinants that stimulate the production of neutralizing antibodies and cellular immunity and provide animal protection on challenge.

Rabies virus glycoprotein	Hepatitis B surface antigen
Vesicular stomatitis glycoprotein	Influenza virus hemagglutinin
Herpes simplex virus glycoprotein D	HIV envelope protein

to immunize mice. The sera from the mice allowed the identification of proteins that are surface-exposed in meningococci, that are conserved across a range of strains, and that induce a bactericidal antibody response.

Recombinant viruses also can be used as vectors to express heterologous antigens and thus function as live vaccines (Table 3.5). The first animal virus to be exploited in this way was vaccinia which had been used previously as a non-recombinant vaccine providing cross-protection against variola virus (smallpox). Vaccinia is particularly attractive as a vaccine carrier since it is very stable when freeze dried, can be produced cheaply, and is easily administered by simple dermal abrasion. The most successful recombinant-virus vaccination campaign to date involved the use of recombinant vaccinia virus expressing rabies-virus glycoprotein. This was administered to the wild population of foxes in central Europe by providing bait spiked with the virus. The epidemiologic impact was startling. Two decades of rabies in Switzerland came to an abrupt end after three vaccination campaigns and the disease was all but eliminated in Belgium. Although vaccinia can be easily converted into novel vaccine candidates, it suffers from the disadvantage that it generates an unacceptably high risk of adverse reactions. For this reason a number of other classes of virus, e.g. adenoviruses and alphaviruses, have been developed as vectors.

Despite the advances in recombinant vaccine development described above, very few recombinant vaccines are approved for use in humans. Of those that are, the most widely used is that affording protection against hepatitis B virus. This consists of the hepatitis B surface antigen expressed in yeast where it forms virus-like particles. Because neither the yeast nor the virus-like particles can replicate in humans, the vaccine is a poor immunogen. Nevertheless, cost–benefit analysis suggests that vaccination against hepatitis B is worthwhile in individuals at risk from contacting the disease. The greatly increased risk of bioterrorism since September 11, 2001, coupled with the transfer to rogue nations of some of the bioweapons developed by the former Soviet Union, almost certainly has prompted a re-think about the development and use of recombinant vaccines. The death of a small number of individuals due to adverse reactions to vaccinia virus is certainly preferable to large scale infection of populations with highly infectious and potentially lethal pathogens!

The immune system generates antibodies in response to the recognition of proteins and other large molecules carried by the pathogens. In the examples cited above, the functional component of the vaccine introduced into the host is the

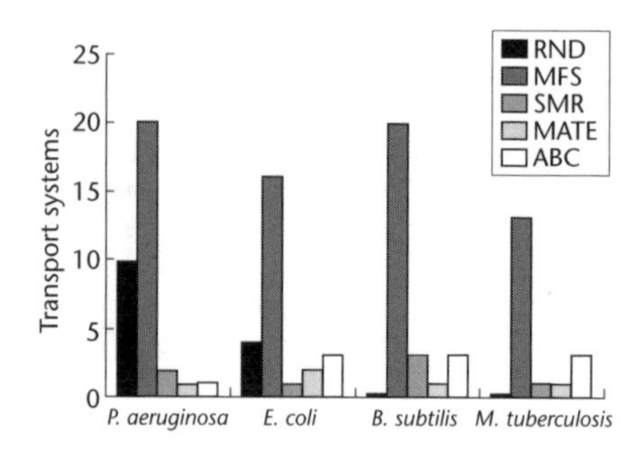

Fig. 3.5 Comparison of the number of predicted drug efflux systems in different bacteria. ABC, ATP-binding cassette family; MATE, multidrug and toxic compound extrusion family; MFS, major facilitator superfamily; RND, resistance-nodulation-cell division family; SMR, small multidrug resistance family.

protein responsible for the elicitation of the immune response. The introduction of DNA into animals does not generate an immune response against the DNA molecule. However, if the DNA is expressed to yield a protein novel to the host then that protein can stimulate the immune system. This is the basis of DNA vaccination (see also p. 194). DNA vaccines generally comprise a bacterial plasmid carrying a gene encoding the appropriate antigen under the control of a strong promoter that is recognized by the host cell. The advantages of the method include its simplicity, its wide applicability, the ease with which large quantities of vaccine can be produced, the ability to include bacterial DNA sequences known to stimulate the immune system, and the ability to treat diseases already established as chronic infections.

Genomics and the development of new antibacterial agents

Microbial resistance to antibiotics is a major problem in human health care. There are two types of resistance: acquired and innate. Acquired resistance arises by mutation or by acquisition of a plasmid carrying genes encoding enzymes that modify antibiotics and render them ineffective. Innate resistance is encoded in the chromosome and the most common form is the presence of multiple genes for antibiotic efflux systems. Whole genome sequencing has shown that these efflux systems are widespread in bacteria and that within one isolate of an organism there can be a whole series of pumps (Fig. 3.5). Because of the prevalence of these efflux pumps, some companies have embarked on programs to find inhibitors of them.

In considering what other cellular functions could be selected as targets for antibiotics, it should be noted that genome analysis has shown that microbial genes can be divided into four types:

1 genes that are conserved in most living organisms;
2 genes that are conserved within a particular phylogenetic kingdom;
3 genes conserved within an order or family;
4 genes that are organism specific.

Products of genes of the first type will not make good targets for antibiotics whereas the other three types offer a degree of selectivity. Analysis of genome sequence data indicates that roughly 25% of bacterial genes are required for growth on a nutrient-rich medium. Of these, 40% have no mammalian counterpart and their gene products are potential targets for new antibiotics. If they are present in most bacteria then they could be used to select broad-spectrum antibiotics, whereas if they are present in only a few bacteria they could be the target of narrow-spectrum antibiotics. For the ultimate in specificity one could target an essential pathway or protein that only occurs in a very restricted group of bacteria. Examples of these are the enzymes involved in lipogenesis and lipolysis and the novel glycine-rich proteins found in *Mycobacterium tuberculosis*. The advantage of the latter approach is that if a suitable antibiotic was identified, its use would be restricted to a small number of clinical indications and this would minimize misuse and the associated development of resistance.

One disadvantage of many antibiotics is that they kill beneficial bacteria as well as harmful ones. A good example is the loss of gut microflora following administration of broad-spectrum antibiotics, an event that can be accompanied by unwanted fungal colonization. Furthermore, because antibiotics are bactericidal there is very strong selective pressure for antibiotic resistance. Both these unwanted effects arise because the antibiotics are targeted at proteins whose function is essential for the growth of the bacteria. An alternative approach would be to target proteins that are essential for virulence but not growth. This would have the added advantage that there is less likelihood that there are mammalian counterparts and hence less chance of problems of toxicity. As discussed earlier, many different virulence determinants have been identified in bacteria, e.g. toxins, adhesins, invasins, etc. However, there probably are many more genes that are essential for virulence which could represent suitable targets for antibiotic therapy. A number of different approaches have been used for identifying these virulence determinants and details are presented below for three of these: *in vivo* expression technology, differential fluorescence induction, and signature-tagged mutagenesis.

In vivo *expression technology*

In vivo expression technology (IVET) was developed to positively select those genes that are induced specifically in a microorganism when it infects an animal or plant host. The basis of the system is a plasmid carrying a promoterless operon fusion of the *lacYZ* genes fused to the *purA* or *thyA* genes downstream of a unique *Bgl*II cloning site (Fig. 3.6). This operon fusion was constructed in a suicide-delivery plasmid. Cloning of pathogen DNA into the *Bgl*II site results in the construction of a pool of transcriptional fusions driven by promoters present in the cloned DNA. The pool of fusions is transferred to a strain of the pathogen carrying a *purA* or *thyA* deletion and selection made for integration into the chromosome. The recombinant pathogen is then used to infect a test animal. Fusions that contain a promoter that is active in the animal allow transcription of the *purA* or *thyA* gene and hence bacterial survival. When the surviving pathogens are reisolated from the test animal, they are tested *in vitro* for their levels of β-galactosidase. Clones that

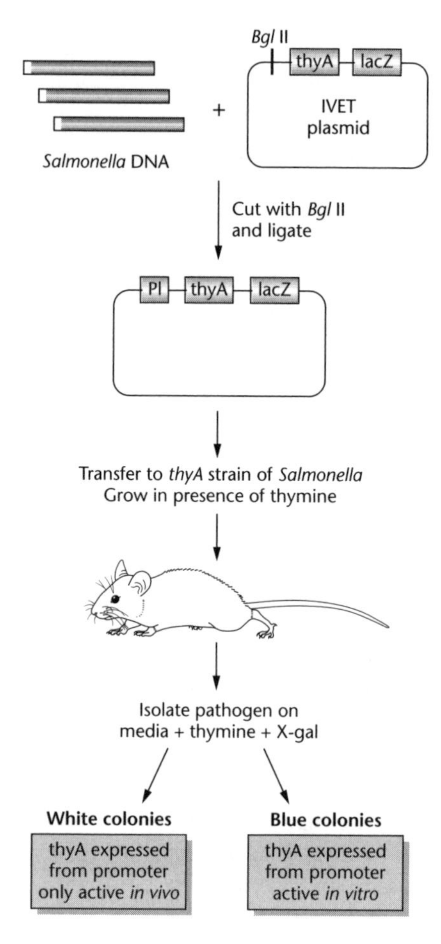

Fig. 3.6 The basic principle of IVET. See text for details.

contain fusions to genes that are **specifically** induced in the test animal show little *lacZ* expression on laboratory media and DNA sequencing can be used to identify the genes to which these promoters belong.

The IVET system was initially developed for use in a murine typhoid model. Since it was first described, IVET has been used with a wide variety of Gram-positive and Gram-negative organisms, including "difficult" organisms, such as *Mycobacterium*.

Differential fluorescence induction

This technique represents a different way of identifying environmentally controlled promoters and was originally developed to facilitate identification of *Salmonella* genes that are differentially expressed with macrophages. Fragments of DNA from *Salmonella* were fused to a promotorless gene for green fluorescent protein (GFP) and returned to the *Salmonella*, which was then used to infect macrophages. Cells that became fluorescent were recovered using a fluorescence-activated cell sorter (FACS) and grown on media in the absence of macrophages. Bacteria that were nonfluorescent in the extracellular environment were sorted and used for a second round of macrophage infection. Bacteria still capable of generating fluorescent macrophages were found to contain GFP fusions that were upregulated by the macrophage's intracellular environment.

Signature-tagged mutagenesis

This technique is a variation on the use of transposon mutagenesis that has been applied to the identification of bacterial virulence genes. The basic principle is to create a large number of different transposon-generated mutants of a pathogenic organism and to identify those mutants that can survive *in vitro* but not *in vivo* – it is a whole-genome scan for habitat-specific genes. Although one could test each mutant individually, this would be very laborious and would use large numbers of animals. By using signature-tagged mutagenesis, one can test large numbers of different mutants *simultaneously in the same animal*. This is achieved by tagging each transposon mutant with a different DNA sequence or tag.

The use of sequence-tagged mutagenesis was first described for the identification of virulence genes from *S. typhimurium* in a mouse model of typhoid fever as shown in Fig. 3.7. The tags comprise different sequences of 40 bp $[NK]_{20}$, where N = A, C, G, or T, and K = G or T. The arms were designed so that the amplification of the tags by PCR with specific primers would produce probes with 10 times more label in the central region than in each arm. The double-stranded tags were ligated into a Tn5 transposon and transferred from *E. coli* to *S. typhimurium* by conjugation. A library of over 1500 exconjugants resulting from transposition events was stored in microtiter dishes. Of these exconjugants, 1152 were selected and prepared as 12 pools of 96 mutants. Each pool was injected into the peritoneum of a different mouse and infection allowed to proceed. Bacteria were then recovered from each mouse by plating spleen homogenates on culture media. DNA was extracted from the recovered bacteria and the tags in this DNA were amplified by PCR. Those tags present in the initial pool of bacteria but missing from the recovered media represent mutations in genes essential for virulence. In this way 28 different mutants with attenuated virulence were identified, some representing previously uncharacterized genes.

The principle of signature-tagged mutagenesis has been extended to the analysis of pathogenicity determinants in a wide range of bacteria and to fungi.

Combating fungal infections

Fungi are eukaryotes and hence are more closely related to humans than bacteria. Indeed, some protein components of fungal and human cells are functionally interchangeable. Such proteins include those involved in the cell division cycle, stress responses, gene regulation, protein localization, metabolism, and ATP synthesis. An important consequence is that it is hard to identify the significant differences between fungal and human cells that might make useful targets for antifungal

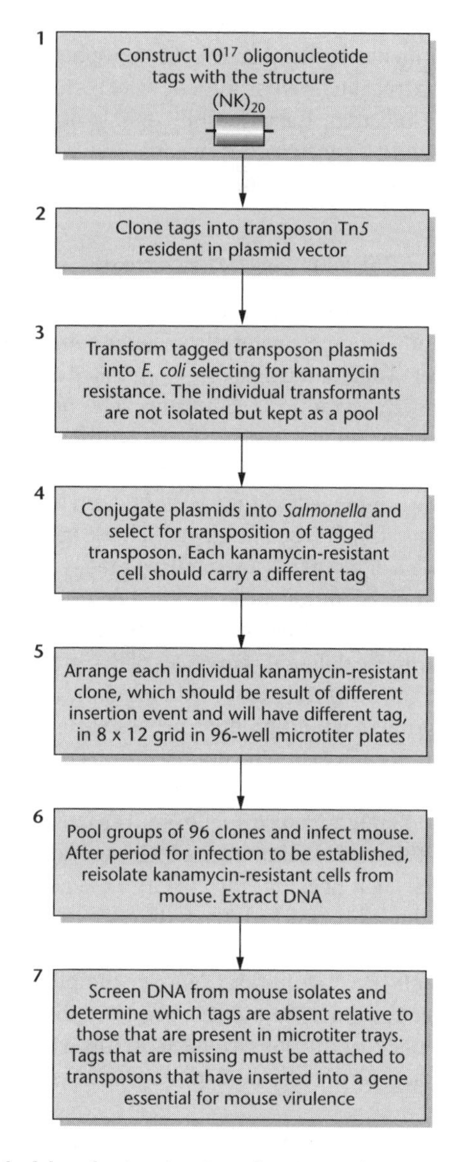

Fig. 3.7 The basic methodology for signature tagged mutagenesis.

drugs. Nevertheless, fungi use similar virulence factors (Table 3.6) to those described earlier for bacteria (see p. 70) and the strategies described earlier for combating bacterial disease are equally applicable to fungal diseases.

Progress in tackling protozoan diseases

As noted earlier (see p. 63, little effort has gone into the development of anti-protozoan drugs despite the fact that they are responsible for the greatest number

Table 3.6 Some virulence factors found in fungi.

Virulence factor	Examples
Toxin production	*Aspergillus fumigatus* gliotoxin
Production of cell adhesins	*Candida* agglutinin-like sequences, integrin-like protein and Hwp1 protein that cross-links fungal cells to host cells
Invasion	Secretion of hydrolytic enzymes by *Candida* and adaptation to acidic pH (vagina)
Evasion of host defenses	Modulation of host immune responses. Switching from mycelial form to yeast-like morphology

Table 3.7 World Health Organization statistics for several protozoan diseases.

Disease	Cases worldwide	People at risk
Malaria	300–500 million	2400 million
Leishmaniasis	12 million	350 million
Chagas disease (American trypanosomiasis)	18 million	100 million
Sleeping sickness (African trypanosomiasis)	300–500,000	60 million

of infections in the world (Table 3.7) and one of them (malaria) alone is responsible for 3 million deaths per year! This lack of effort is illustrated by the fact that of the 1400 new drugs developed in the last 25 years, less than 1% was for the treatment of protozoan diseases. Part of the problem is that protozoan diseases are seen as unique to underdeveloped countries. However, malaria was rife in the USA and Europe, including the UK, in the late nineteenth and early twentieth centuries and may return if global warming persists. Another reason for the lack of effort has been an inadequate knowledge of parasite biochemistry. However, good progress is being made in the sequencing of the genomes of a number of protozoan pathogens and the information being generated is providing new opportunities for combating these diseases.

In tackling any infectious disease there is a choice between disease prevention (vaccination) and disease treatment (antibiotics). For diseases prevalent in developing countries the former approach is undoubtedly preferable since it puts the least burden on rudimentary health care systems. In developing vaccines for protozoan diseases, malaria represents the greatest challenge. First, the disease goes through different stages in the human body (Box 3.5) and at each stage the parasite expresses different antigens. Thus a vaccine effective in killing liver-stage parasites may not inhibit the growth of blood-stage parasites. Second, a single parasite clone contains roughly 50 different copies of the gene for the variable surface antigen. During a chronic infection, each successive wave of parasitemia expresses a new variant surface antigen thereby allowing parasite multiplication despite the presence of antibodies directed against the preceding parasitic wave. Finally,

Box 3.5 The malarial life cycle

When an infected mosquito bites a human it injects malarial sporozoites into the bloodstream. These sporozoites find their way via the blood to the liver where they infect hepatocytes. After 8–25 days, depending on the malarial species, sporozoites mature to infectious tissue schizonts . These schizonts release 20,000–30,000 merozoites into the bloodstream where they proceed to infect erythrocytes. The asexual replication cycle then continues in the erythrocytes every 48–72 hours until treatment, development of immunity or death occurs. The synchronous release of mature merozoites from erythrocytes coincides with the abrupt onset of periodic fever and shaking chills that characterize the disease. Some merozoites develop into male or female gametocytes but conversion to gametes and subsequent fertilization does not occur until the gametocytes are acquired by a mosquito when it takes a blood meal. In the mosquito, the zygote develops into an oocyst that eventually ruptures to release sporozoites that move to the salivary glands in readiness for transmission to a new human host.

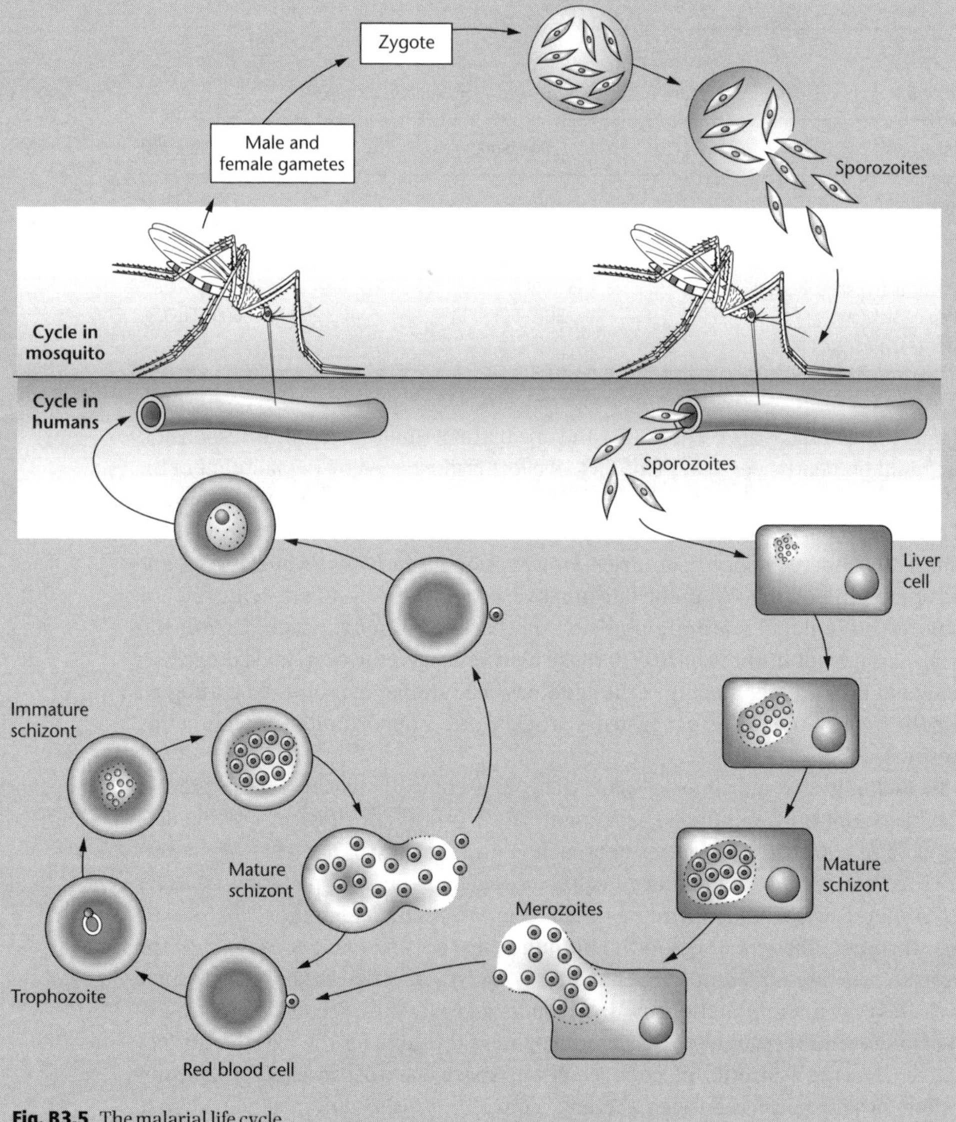

Fig. B3.5 The malarial life cycle.

although animal models exist (*Plasmodium yoelii* in rodents) there is no good model of human malaria. However, now that the genome of *P. falciparum* has been completely sequenced there is hope that new vaccine candidates can be identified in a manner analogous to that described earlier (see p. 76) for bacteria. Already a large number of proteins that are likely to be surface expressed have been identified. The challenge now is to identify those that have limited antigenic diversity and are expressed in more than one stage of the life cycle.

Progress in developing new antibiotics for treating protozoan diseases has been facilitated by the sequencing of the genome of an organelle unique to some of them. This organelle is the apicoplast and among the human pathogens it is found in *Plasmodium* (malarial parasite), *Toxoplasma*, and *Cryptosporidium*. It is a chloroplast-like plastid and, not surprisingly, sequence analysis has shown that many of the genes encoded in its genome have prokaryotic characteristics. This explains the susceptibility of apicomplexan parasites to antibiotics that interfere with bacterial transcription and translation, e.g. rifampicin and doxycycline. A number of apicoplast metabolic pathways have been identified from sequence data and these too are characteristic of bacteria and plants rather than mammalian cells. As such, they are providing new targets for drug development.

Isoprenoids are essential lipids that include molecules such as cholesterol and coenzyme Q (ubiquinone). They are synthesized by condensation of isopentyl diphosphate. In mammals, isopentyl diphosphate is synthesized from mevalonate whereas in bacteria and plants it is synthesized via 1-deoxy-D-xylulose 5-phosphate (DOXP). Genomic analysis revealed that the apicoplast genome encodes two key enzymes of the DOXP pathway: DOXP synthase and DOXP reductoisomerase. The antibiotic fosidomycin inhibits DOXP reductoisomerase and hence should be effective against the malarial parasite, whereas inhibitors of the mevalonate pathway should have no effect. This is indeed the case. The apicomplexan genome also encodes a type II fatty acid synthase whereas mammalian cells use a type I synthase. Although the former is susceptible to inhibition by triclosan the latter is not. Crucially, administration of triclosan to malaria-infected mice results in the complete clearance of the malarial parasites with no adverse effect on the mammalian host.

It is unfortunate that trypanosomes and *Leishmania* do not harbor an apicoplast-like organelle to target for chemotherapy but, hopefully, genome sequencing will identify potential targets. In the case of African trypanosomiasis, genomics may provide an alternative approach. Sleeping sickness caused by *Trypanosoma brucei* is spread by the bloodsucking tsetse fly. The fly is dependent for its nutrition and fertility on an obligate endosymbiotic bacterium, *Wigglesworthia glossinidia*. This bacterium is related to *Escherichia coli* although its genome is only about one-seventh the size. The genome of *W. glossinidia* encodes over 60 proteins involved in the synthesis of vitamins that the fly does not get from its restricted diet of blood. By targeting either these gene products or other *Wigglesworthia* proteins it may be possible to selectively kill off the tsetse fly.

Another bacterium that is essential for reproduction in insects is *Wolbachia*. A serendipitous discovery from the genome project on *Onchocerca volvulus*, the filarial worm that causes river blindness, was the identification of *Wolbachia* genome sequences. Infection with *Wolbachia* appears to be essential for *Onchocerca* survival and this may permit the development of new drugs for treating filarial diseases.

Table 3.8 The Baltimore classification of viruses.

Class	Characteristics
Class 1	Viruses have double-stranded (dsDNA) genomes. DNA replication and transcription occur by processes similar to that used by the host cell. Transcription can be from either strand of DNA
Class 2	Viruses have single-stranded (ssDNA) genomes. The DNA can be of positive or negative sense. DNA must be converted to a ds form before the synthesis of mRNA can proceed
Class 3	Viruses have dsRNA genomes but transcription only from one strand. Transcription from an RNA template requires RNA-dependent RNA polymerase and this is viral-encoded and packaged with the virus particle
Class 4	Viruses have an ssRNA genome of the same sense (i.e. positive sense) as mRNA and which can be translated. Replication is via a dsRNA intermediate
Class 5	Viruses have an ssRNA genome which is complementary in sequence to their mRNA (negative sense) and are known as negative-strand viruses. Synthesis of mRNA occurs by transcription of the genome strand and involves virus-encoded enzymes that are packaged in the virus particle
Class 6	Viruses have a positive-sense ssRNA genome but replicate via a dsDNA intermediate that is synthesized by a virus-coded reverse transcriptase that is packaged in the virus particle
Class 7	Viruses have a dsDNA genome but replicate via a positive-sense RNA intermediate before being converted back to a dsDNA form by reverse transcriptase. These viruses are called "reversiviruses"

Developing antiviral drugs

Effective vaccines have facilitated the elimination or control of a number of important viral pathogens such as smallpox, measles, mumps, and rubella. However, other viral diseases have proved intractable to the vaccine approach and these include HIV, hepatitis C virus, viruses infecting the respiratory tract, human papilloma viruses, herpes viruses, and the many hemorrhagic fever viruses. Where vaccines are not available, the only alternative is the use of antiviral drugs. Currently, over 30 antiviral drugs have been licenced but most of these are of use against a very narrow spectrum of viruses and 50% are only for the treatment of HIV infection. This is a direct consequence of their mode of action (see below). Nor are many of these drugs well tolerated and the reason for this is that the targets for these drugs are little different in viruses and their hosts.

Many of the drugs currently on sale were selected by the classical method of empiric screening of compounds for antiviral effects. Today, this empiric approach has been replaced by rational design of inhibitors for particular targets and relies heavily on a detailed understanding of the molecular biology of virus multiplication. In theory, these targets could be from any of the key steps in viral infection: attachment, replication, transcription, translation, assembly, or release. In practice, most are targeted at viral DNA synthesis, virus polyprotein cleavage, and virus release from the host cell.

Although there are many different types of virus they can be classified into just seven types based on their genome structure and their modes of replication and gene expression (Table 3.8). This simplification, which was devised originally by the Nobel laureate David Baltimore, facilitates an understanding of the limited

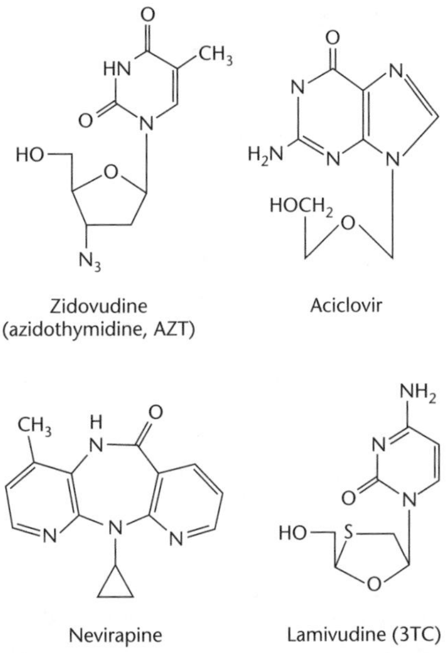

Zidovudine
(azidothymidine, AZT)

Aciclovir

Nevirapine

Lamivudine (3TC)

Fig. 3.8 Structures of some antiviral compounds.

spectrum of activity of some of the key antiviral drugs. Two examples are given below.

Herpesviruses (class 1) have double-stranded DNA genomes that are replicated by viral-encoded DNA polymerases. A number of nucleoside analogs such as aciclovir, ganciclovir, and famciclovir are available for the treatment of herpesvirus infections and act as premature chain terminators. All of these compounds target the viral DNA polymerase but, before they can interact with viral DNA synthesis, they need to be phosphorylated intracellularly to the triphosphate form (see Fig. 8.7). The first phosphorylation step is mediated only by the herpes simplex virus- and varicella zoster virus-encoded thymidine kinase or the cytomegalovirus-encoded protein kinase. In this way the activity of these therapeutic agents is confined to virus-infected cells but their antiviral spectrum is restricted to those viruses that encode an enzyme capable of phosphorylating deoxyribonucleoside analogs.

Retroviruses such as HIV belong to class 6 and hence reverse transcriptase activity is essential for viral replication. Since reverse transcriptase does not normally occur in mammalian cells it represents an attractive target for antiviral drug design. Two types of reverse transcriptase-based inhibitors have been developed. The first of these are dideoxynucleosides such as azidothymidine (AZT) that act as substrates for the reverse transcriptase but cause premature chain termination. These are known as nucleoside analog reverse transcriptase inhibitors (NRTIs). The second class of inhibitors is the non-nucleoside reverse transcriptase inhibitors (NNRTIs). These compounds interact with an allosteric binding site that is present only in HIV-1 reverse transcriptase. Examples are shown in Fig. 3.8.

Box 3.6 Highly active antiretroviral therapy (HAART) in the treatment of AIDS

Development of viral resistance to antiretroviral drugs used for the treatment of HIV infection is a major cause of treatment failure and limits options for alternative treatment regimens. To prevent the development of resistance, the current practice is to treat HIV-infected individuals using a combination of at least three drugs. This is known as highly active antiretroviral therapy (HAART). Generally, treatment combinations comprise two NTRIs plus a PI or an NNRTI, although sometimes three NTRIs may be used. The rationale for HAART is set out below.

The virus population in a person infected with an RNA virus such as HIV has been termed a quasi species because of the existence of genetically distinct viral variants that evolve from the initial virus inoculum. The variants are generated because RNA polymerases do not have the proofreading function found in DNA polymerases. Thus, as single-stranded RNA viruses replicate, each newly copied genome differs from the parental virus by an average of one nucleotide. An estimated 10 billion HIV virions are produced daily in an established HIV infection. If each contains on average one mutation per 9200-nucleotide genome, replication-competent virus with every single drug-resistance mutation is likely to be generated daily. Double mutants are less likely, and the probability of three or more drug-resistance mutations in the same genome is very low.

For some antiretroviral drugs such as lamivudine (an NRTI) and the NNRTIs, a single mutation in the HIV can confer high-level resistance. For some other drugs, such as zidovudine (an NRTI) and certain protease inhibitors, high-level resistance requires accumulation of three or more resistance mutations in a single viral genome. From clinical practice with HAART, two things are known. First, the higher the plasma concentrations of a protease inhibitor the more slowly resistance mutations emerge. Second, the lower the plasma levels of HIV RNA the longer it takes for drug failure to occur. In patients with suppression of plasma HIV RNA to below 50 copies per milliliter for 1 year no resistance mutations can be discerned even though replication-competent virus persists. Clearly, drugs for which only a single mutation is required for high-level resistance should be reserved for special cases and not used in the initial combination therapy.

Novel methods for combatting AIDS are described on page 193.

Viral proteases are crucial in the life cycle of many viruses including HIV, herpesviruses, and rhinoviruses (common cold). These proteases cleave newly expressed precursor polyproteins into smaller, functional and structural viral proteins. Because of their key role in viral multiplication they have been favored as targets for antiviral drugs and a number of protease inhibitors (PIs) are licensed for the treatment of HIV infections (Box 3.6).

As noted earlier, viruses use many of the same enzymes as their mammalian hosts thereby making the development of nontoxic antiviral drugs very difficult. One approach that avoids this problem is the use of antisense drugs (see also p. 190). These drugs are short oligoucleotides complementary to a specific region of a viral-encoded mRNA. By binding to the mRNA to form a double-stranded structure they specifically inhibit translation of viral-encoded proteins. Because short oligonucleotides would be readily degraded inside host cells, antisense drugs usually are synthesized from nucleoside analogs that are resistant to nuclease attack. To date, only one antisense drug has been licenced for sale and that is one targeted at retinitis caused by cytomegalovirus. Treatment is by direct injection of the drug into the eyeball and this highlights one of the problems of antisense drugs – getting them to their target.

Further reading

POGM: Chapter 14 has a section on the use of combinatorial biosynthesis to generate new antibiotics.

POGA: Chapter 4 describes in detail the different methods of detecting physical markers in genomes that are used by microbiologists for typing bacterial and viral strains.

Atlas RM (2002) Bioterrorism: from threat to reality. *Annu Rev Microbiol* **56**, 167–185.
This review covers everything you did not want to know about bioterrorism.

Bonten MJM, Willems R, Weinstein RA (2001) Vancomycin-resistant enterococci: why are they here, and where do they come from? *Lancet Infect Dis* 1, 314–325.
A fascinating analysis of the epidemiology of VREs.

De Clercq E (2002) Strategies in the design of antiviral drugs. *Nature Rev Drug Discovery* **1**, 13–25.
This article contains everything you wanted to know about antiviral drugs.

Finlay RB, Falkow S (1997) Common themes in microbial pathogenicity revisited. *Microbiol Mol Biol Rev* **61**, 136–169.
Joyce EA, Chan K, Salama NR, Falkow S (2002) Redefining bacterial populations: a post-genomic reformation. *Nature Rev Genet* **3**, 462–473.
These two reviews give an excellent insight into the mechanisms of bacterial pathogenicity, albeit from different perspectives.

Grandi G (2001) Antibacterial vaccine design using genomics and proteomics. *Trends Biotechnol* **19**, 181–188.

Hiramatsu K, Cui L, Kuroda M, Ito T (2001) The emergence and evolution of methicillin-resistant *Staphylococcus aureus*. *Trends Microbiol* **9**, 486–493.
This paper presents a very detailed analysis of the genetic structure and origin of MRSAs.

Gardner MJ, Hall N, Fung E *et al* (2002) The genome sequence of the human malaria parasite *Plasmodium falciparum*. *Nature* **419**, 498–511.
This is just one article from a collection of papers on malaria genomics and proteomics that appeared in the same journal issue.

Karlin S (2001) Detecting anomalous gene clusters and pathogenicity islands in diverse bacterial genomes. *Trends Microbiol* **9**, 335–343.
A detailed, but mathematical, review of the methods for identifying PAIs.

Loferer H (2000) Mining bacterial genomes for antimicrobial targets. *Mol Med Today* **6**, 470–474.

Mecsas J (2002) Use of signature-tagged mutagenesis in pathogenesis studies. *Curr Opin Microbiol* **5**, 33–37.
An excellent review of this very powerful tool.

Schoolnik GK (2002) Functional and comparative genomics of pathogenic bacteria. *Curr Opin Microbiol* **5**, 20–26.

CHAPTER FOUR

Analyzing and treating genetic diseases

Genetic disease in context

All known genetic diseases can be classified into one of three types: single gene disorders, chromosome abnormalities, and polygenic diseases. Single gene disorders are caused by mutations in a single gene on an autosome, a sex chromosome, or in mitochondrial DNA. Regardless of whether the mutations are dominant or recessive, they exhibit obvious and characteristic pedigree patterns. Although any one particular single gene disorder may be rare in the population as a whole, collectively they affect 2% of the population at some stage of their life. The incidence of serious single gene defects in young children is 0.36% but rises to 6% among hospitalized children.

Chromosome abnormalities take a number of forms. These include an excess or deficiency of either part or all of a chromosome or the translocation of part of a chromosome to a new site. Such abnormalities occur in 0.7% of live births but the frequency rises to 50% in spontaneous abortions occurring during the first 3 months of pregnancy. Chromosome abnormalities are very common in cancer cells but they result from clonal propagation of somatic cell mutations. They are not inherited abnormalities, however the susceptibility to their formation may be inherited.

Until recently, polygenic disorders often were not recognized as genetic diseases because they do not show characteristic pedigree patterns and because severity can be influenced by lifestyle factors. Even when they were recognized as having a genetic component, they were poorly understood. Polygenic disorders are not due to a single mutation in one gene. Rather, they are the result of small variations in a number of genes that together can predispose an individual to a serious defect. The frequency of polygenic disease is difficult to determine but current estimates range from 5% in children to over 60% in the total population with the greatest impact being on the elderly.

Today, there is little that can be done for individuals afflicted with chromosome abnormalities. However, recent advances in biotechnology and genomics are providing a range of exciting therapies for patients with single gene defects. These advances include the provision of therapeutic proteins, antisense technology, gene therapy, and gene repair. The development of novel molecular diagnostic procedures also is permitting improved prenatal diagnoses but these can lead to difficult ethical decisions. In the last 5 years there has been a tremendous increase in our understanding of polygenic disorders. It is too soon for this knowledge to have affected clinical practice but one outcome could be the better matching of drugs to disease phenotypes in what has become known as **personalized medicine**. It also will transform the way in which new drugs are tested in clinical trials.

A key driver of advances in the treatment of genetic diseases will be the continuing analysis of information deriving from the complete sequencing of the human genome and the genomes of related species such as the mouse. This information should provide new insights into many diseases and promote the development of far better diagnostic tools, preventative measures, and therapeutic methods.

Detecting single gene disorders

Mutations can be of two types: point mutations caused by nucleotide substitutions and mutations involving larger stretches of nucleotides, i.e. insertions and deletions. Examples of these are shown in Table 4.1. It should be noted that some

Table 4.1 Examples of single gene disorders.

Genetic change	Example
Point mutation	
Missense mutation resulting in amino acid substitution in protein which affects protein function	Change of A to T in the sixth codon of the β-globin gene converts a glutamate residue to valine and results in sickle cell anemia
Nonsense mutation caused by generation of premature stop codon leading to formation of truncated protein	Mutation in codon 39 of the β-globin gene converts a glutamine codon to a stop codon resulting in β^0-thalassemia
RNA processing mutation leading to abnormal splicing	RNA splicing mutants are the commonest cause of β-thalassemia
Regulatory mutation affecting gene expression, e.g. transcription factor binding	A change of G to A upstream of the promoter of the γ-globin gene results in hereditary persistence of fetal hemoglobin
Insertion or deletion	
Addition or deletion of small number of bases (3 or multiple of 3 bases) without causing frameshift	Three base deletion in the cystic fibrosis gene removes phenylalanine residue; accounts for most cases of cystic fibrosis in Caucasians
Addition or deletion of small number of bases resulting in frameshift	Four base insertion in the hexosaminidase A gene in Ashkenazic Jews causes Tay–Sachs disease
Insertion of dispersed repeated sequence such as Line or Alu element	
Expansion of trinucleotide repeat sequence	More than 35 copies of the CAG repeat in the huntingtin gene result in Huntington disease in adults

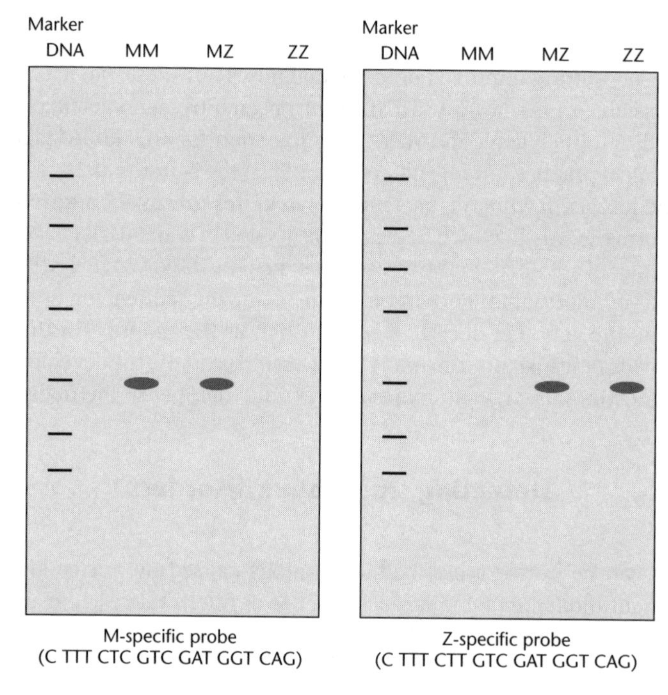

Fig. 4.1 Schematic representation of the use of oligonucleotide probes to detect the normal α_1-antitrypsin gene (M) and its Z variant. Human DNA obtained from normal (MM), heterozygous (MZ) and homozygous variant (ZZ) subjects is digested with a restriction endonuclease, electrophoresed, and fragments Southern blotted on to a nylon membrane. The pattterns shown were obtained on autoradiography of the filter following hybridization with either the normal (M-specific) or variant (Z-specific) probe.

deletions extend into contiguous genes and, strictly, the diseases that they cause are not single gene disorders.

Not all single nucleotide changes within a gene result in a phenotypic effect (genetic disease). This could be because the change does not alter the primary amino acid sequence of a protein or, even if it does, the resulting change does not alter the functional properties of the protein. Nevertheless, the mutated gene is different genetically and the change is known as a single nucleotide polymorphism (SNP, pronounced "snip"). SNPs also occur in noncoding regions of the genome and, as will be seen later, are very useful in diagnosing genetic disease and in investigating polygenic disorders.

There are many different methods for detecting single gene disorders and all rely to a greater or lesser extent on the hybridization of allele-specific oligonucleotides. The earliest molecular methods involved electrophoretic separation of human DNA digested with a restriction endonuclease followed by hybridization of test probes using Southern blotting (see Chapter 1 for an overview). An example of this technique is the detection of a point mutation in the α_1-antitrypsin gene that is responsible for hereditary emphysema. Two different 19-mer oligonucleotides are used,

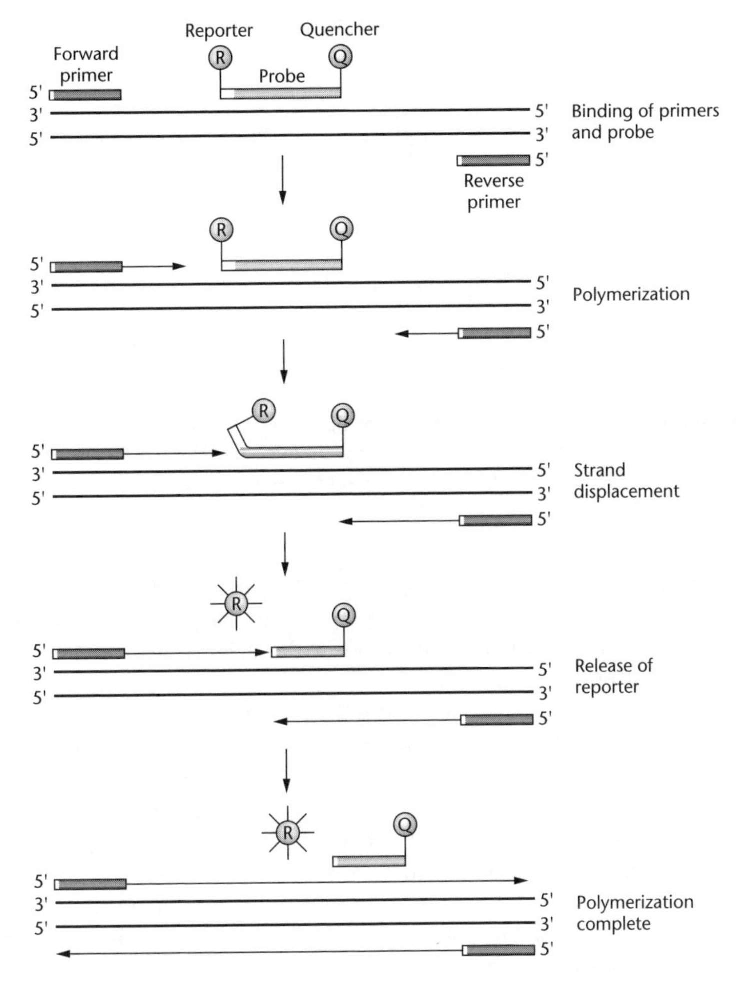

Fig. 4.2 The TaqMan assay (see text for details).

one that is complementary to the normal allele and the other that is complementary to the mutant allele. These provide sufficient discrimination (Fig. 4.1) to distinguish normal and affected individuals and heterozygotes (carriers); other examples are shown on page 12.

Methods involving solid matrices like the one just described have the advantage that they give a "pictorial" result that is easy to interpret. The disadvantage is that they involve electrophoresis, as well as Southern blotting, which are both time consuming and messy. Fortunately, a number of mutation detection assays have been developed that work entirely in solution. The most popular of these assays is the TaqMan assay (Fig. 4.2). In this method, a 100-bp region flanking the polymorphism is amplified by the PCR in the presence of two probes, each specific for one or other allele. The probes have a reporter fluor at the 5′ end but do not fluoresce when free in solution because they have a quencher at the 3′ end. During the PCR, the *Taq*

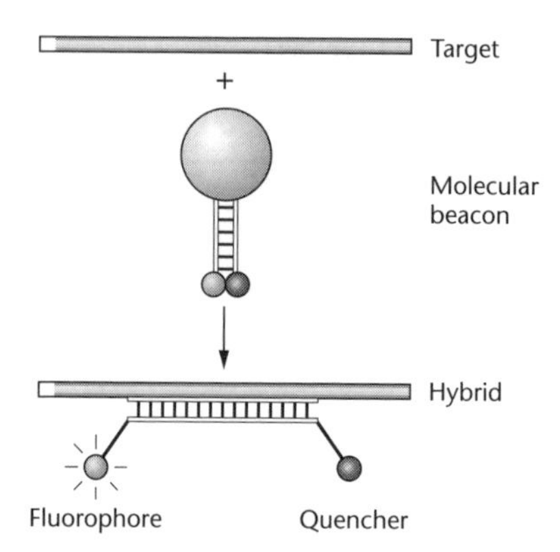

Fig. 4.3 Principle of molecular beacons. On their own, these molecules are nonfluorescent, because the stem hybrid keeps the fluorophore close to the quencher. When the probe sequence in the loop hybridizes to its target, forming a rigid double helix, a conformational reorganization occurs that separates the quencher from the fluorophore, restoring fluorescence.

DNA polymerase encounters a probe specifically base paired with its target and liberates the fluor, thereby increasing net fluorescence. The presence of two probes, each labeled with a different fluor, allows one to detect both alleles in a single tube without any post-PCR processing.

An alternative to the TaqMan assay is the use of molecular beacons. These are oligonucleotide probes having two complementary sequences flanking the target-specific sequence and a reporter fluor and quencher dye at opposite ends (Fig. 4.3). After PCR amplification of the target DNA, the molecular beacons are added. If hybridization occurs to the target DNA, the two dyes are separated and the reporter fluor emits a signal. If hybridization does not occur the quencher prevents fluorescence of the reporter dye. Molecular beacons can be made in many different colors by using a broad range of fluorophores and this enables multiple mutations to be analyzed simultaneously.

The above methods are very useful where it is known that the genetic disease is caused by just one or a few different mutations. Where many different mutations are known in a gene, e.g. the β-globin gene, other methods may have to be used. If it is known that there can be many different deletions in the gene of interest then western blotting may be of value. In this method, proteins isolated from a cell extract are separated according to size by polyacrylamide gel electrophoresis and then transferred to a membrane. The membrane is incubated with antibodies that specifically recognize the protein to be analyzed. The specific interaction between the antibody and its antigen is detected by the addition of a second antibody against the first that carries a histochemical or fluorescent tag. An example of the use of this methodology in the analysis of patients with muscular dystrophy is shown in Fig. 4.4.

The detection of variable number tandem repeats, such as trinucleotide expansions, has to be done using Southern blotting. In this case, one looks for changes in the size of restriction fragments carrying the region of interest (Fig. 4.5).

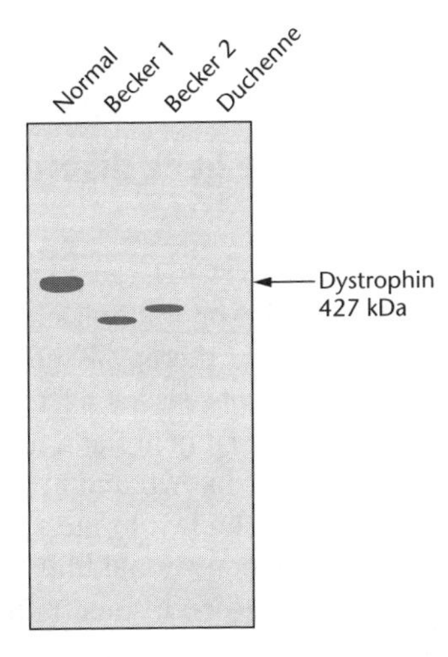

Fig. 4.4 A western blot demonstrating the presence or absence of the muscle protein dystrophin (arrow) in protein extracts from patients with the severe Duchenne or mild Becker form of X-linked muscular dystrophy.

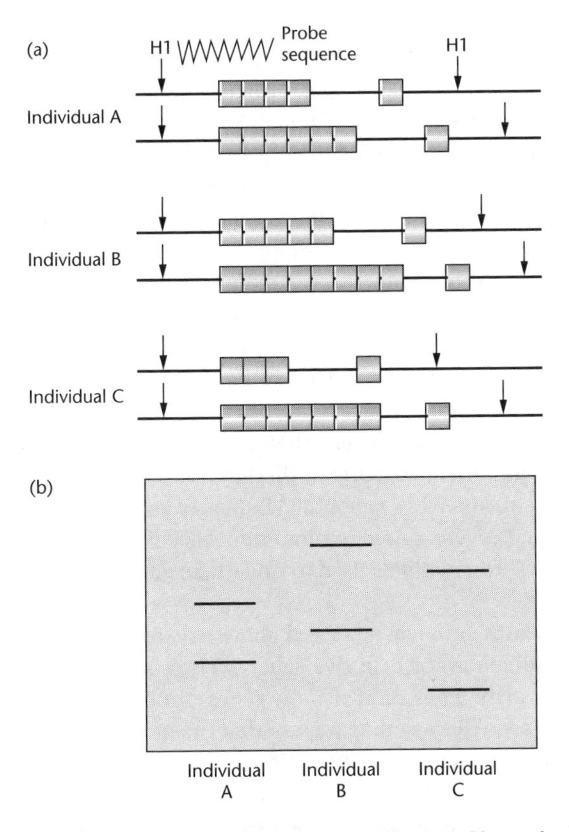

Fig. 4.5 Restriction fragment length polymorphisms caused by a variable number of tandem repeats between the two *Hin*FI restrictions sites. (a) The DNA structure for three different individuals. (b) The pattern obtained on electrophoresis of *Hin*FI cut DNA from the three individuals after hybridization with a probe complementary to the sequence shown by the wavy line in (a).

Treating single gene disorders

Today, treatment of monogenic disorders is largely based on replacing the defective protein, improving its function, or minimizing the consequences of its deficiency. Unfortunately, fully effective therapy is not possible for more than 80% of disorders. There are a number of reasons for this unsatisfactory state of affairs. First, research has shown that treatment has a much higher probability of success if the basic biochemical defect is known and its pathogenesis fully understood. Currently, the gene defect is not known in most single gene disorders, far less its pathophysiology, but this situation should improve greatly now that the human genome has been sequenced. Second, some mutations exert their effect in the fetus and by childbirth it is too late to begin treatment. This underscores the benefit of prenatal diagnosis since it may be possible to begin treatment *in utero*. For example, the effects of biotinidase or methylmalonic aciduria can be counteracted by the administration of biotin and cobalamin respectively during pregnancy.

The different strategies that can be employed in treating genetic disease are summarized in Fig. 4.6. As can be seen from the figure, some of the therapies are rooted in biotechnology. Of these, the best known is protein replacement and examples of the use of recombinant therapeutic proteins to treat genetic disorders are given in Table 4.2. The production of these proteins is covered in Chapter 6. It should be noted that a major disadvantage of therapeutic proteins is that they have to be given by injection and this limits their utility.

An alternative way of using recombinant DNA technology is to undertake correction of single gene disorders by the administration of corrective genes to **somatic** tissues. This is known as gene therapy. Three variants of the technique exist and all are at a very early stage of development. The most advanced of these techniques is the introduction into a tissue of a fully functioning gene that is able to compensate for a mutant cellular gene that has a loss-of-function mutation. This is the generally recognized concept of gene therapy and it has been shown to work in a small number of patients with hemophilia B (lack of factor IX), adenosine deaminase deficiency, and severe combined immunodeficiency disease (SCID, defective cytokine receptor). The methods used to undertake gene therapy are described in detail in Chapter 8.

In the gene therapy protocol described above, treated cells carry the new functional gene in addition to the defective gene(s). There are two disadvantages with this approach. First, the expression of many genes is under complex regulatory control and effective gene therapy may require that the new gene carries the requisite regulatory sequences. Second, many mammalian genes are very large. An example of this is Duchenne muscular dystrophy where the cDNA of the dystrophin gene is 11 kb and much too large to fit inside the current generation of gene therapy vectors. In these cases, a better alternative would be to **correct** the defective gene rather than adding a new, functional gene. This is sometimes known as **gene repair**. In essence, a functioning gene is introduced into cells of the appropriate tissue and repair induced by stimulating homologous recombination. Note that

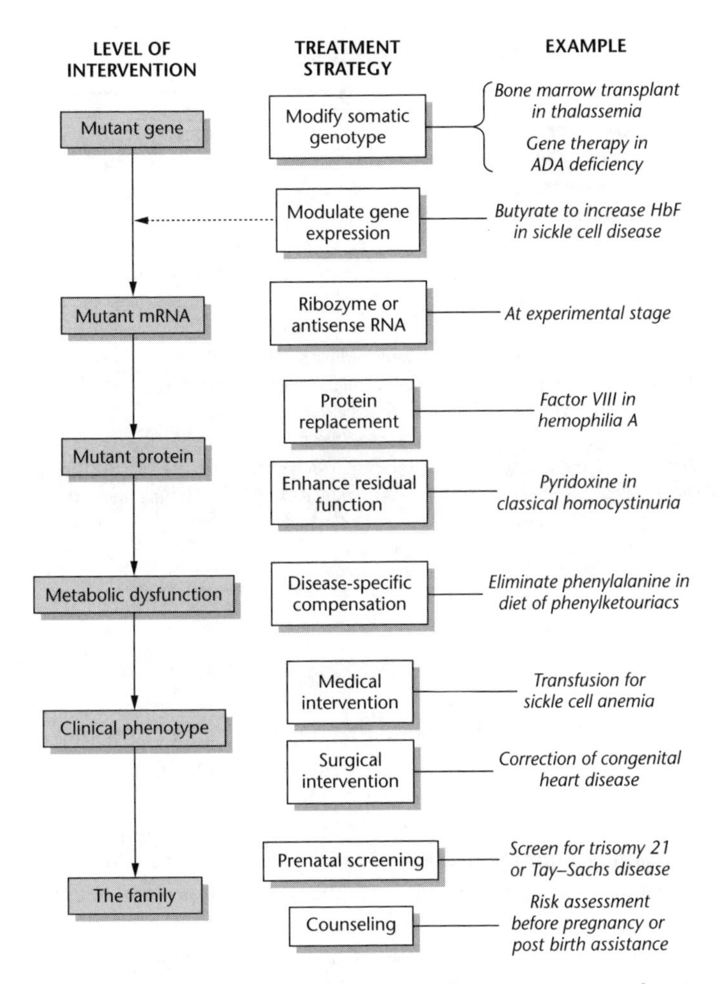

Fig. 4.6 The various methods for treating genetic disease. If butyrate is given to infants it can prevent the postnatal switch from γ-globin to β-globin by an unknown mechanism. In homocystinuria due to cystathione synthase deficiency the cofactor (pyridoxal phosphate) is not linked to the apoenzyme but patients responded to high doses of vitamin B_6 (the precursor of pyridoxal phosphate).

Table 4.2 Some genetic disorders that can be treated with recombinant proteins.

Disease	Therapeutic protein
Pituitary dwarfism	Human growth hormone
Hemophilia A	Factor VIII
Hemophilia B	Factor IX
Hereditary emphysema	α_1-antitrypsin
Gaucher's disease	Glucocerebrosidase
ADA deficiency	Adenosine deaminase

Box 4.1 Ribozymes

Ribozymes are catalytic RNA molecules that can promote specific biochemical reactions without the need for ancillary proteins. The RNA-catalyzed reactions can either be intramolecular or intermolecular. An example of an intramolecular reaction is the splicing out of introns. In intermolecular reactions, other RNA molecules act as substrates and the ribozyme is the catalyst, i.e. the ribozyme acts just like a conventional enzyme and emerges unchanged from each reaction. The best example of a natural ribozyme catalyzing intermolecular reactions is the RNA component of bacterial ribonuclease P, an enzyme involved in the maturation of tRNA molecules. This enzyme consists of a protein–RNA complex but the RNA component alone can cleave tRNA molecules. Indeed, the protein component alone has no RNase activity and its role probably is to prevent electrostatic repulsion between the substrate RNA and the ribozyme, both of which are negatively charged.

Using the techniques of modern molecular biology, ribozymes can be designed as human therapeutics to recognize, bind and digest specific mRNA sequences. The specific binding is achieved by ensuring that the ribozyme has a sequence complementary to the target mRNA. After binding, the enzymatic activity of the ribozyme cleaves the mRNA thus preventing its translation into protein. An example of a therapeutic ribozyme is Herzyme that targets the mRNA for human epidermal growth factor-2 (see Chapter 8). Herzyme is at the stage of clinical trials to determine its effectiveness as an anticancer agent.

gene repair would be appropriate for correcting those genetic diseases where the mutation lies in a regulatory region and results in decreased expression of the gene product.

In some cases, a genetic disorder is due to overexpression of one or more genes as a result of a mutation in a regulatory protein. For example, many cancer cells have mutations in the gene encoding the p53 protein, a DNA-binding transcription factor that acts as a tumor suppressor. Conventional gene therapy could be used to promote the synthesis of functional copies of the regulatory protein. Alternatively, gene therapy could be used to downregulate the translation of the overexpressed proteins. This has been achieved in tissue culture in two ways. In the first of these methods, vectors introduce a gene that is transcribed into antisense RNA (see p. 190). In the second, the introduced gene encodes a ribozyme, an RNA molecule that can cleave mRNA molecules very specifically (Box 4.1).

Finding genes for monogenic diseases and determining gene function

As noted earlier, the effective treatment of genetic diseases requires an understanding of the biochemical defect that is caused by the mutation in question. This in turn requires that the defective gene is isolated and its function determined. Isolating genes demands knowledge of their chromosomal location. So, in practice, one starts with a collection of pedigrees in which the gene of interest is segregating. These families are studied with multiple polymorphic markers until evidence for chromosomal location is obtained. Then linkage to other markers on that chromosome is determined. Clearly, the greater the number of markers available per chromosome, the more useful will be the mapping data. The closer the gene of interest to a known marker, the easier will be the later stages. Because only a limited number of human phenotypic traits (genetic markers) are available and most are widely spaced, geneticists have developed a whole series of molecular markers that are spaced evenly across the genome. SNPs (see p. 23) are the most favored markers because they occur on average once in every 1000 bases. Absolute location of the gene of interest is facilitated if the corresponding trait is associated with a cytogenetically visible chromosome abnormality.

Positional cloning

Once a gene has been mapped to a region of a chromosome, its absolute location and its function are determined by one of two approaches: positional cloning and the candidate gene method. A good example of positional cloning is the isolation and characterization of the muscular dystrophy (DMD) gene. In this instance, isolation of the gene was facilitated by the availability of affected individuals in which the disease was due either to partial gene deletion or translocation. These cytogenetic defects permitted the location of the gene to the p21 region of the X chromosome. The availability of patients with deletions was particularly fortuitous as it enabled subtraction cloning to be undertaken. For this, genomic DNA from a normal individual was digested with a restriction endonuclease that produces cohesive ends. These fragments were denatured and annealed with a 200-fold excess of denatured DNA from a deletion patient and which had been sonicated to produce ragged ends. Under these circumstances, Xp21 DNA present in normal individuals but absent in the deletion patient will anneal with itself and will be the only DNA with cohesive ends suitable for cloning in a vector.

The cloned DNA isolated as described above was used as a probe in Southern blot hybridizations against normal and patient DNA samples to identify those that could be carrying the *DMD* gene. One of six fragments derived from the Xp21 region detected deletions in a proportion of patients with DMD and was tightly linked to the disease in family studies. Approximately 200 kb of contiguous DNA in this region was isolated and the nonrepetitive DNA identified by hybridization to chicken, rodent and primate DNA. The principle of this technique, known as **zoo blotting**, is that coding sequences are conserved during evolution whereas noncoding sequences are not. One particular region hybridized at high stringency to the DNA from the other species and sequencing disclosed an exon bounded by splicing signals. A probe against this region detected a large RNA species in human fetal muscle that was absent from other tissues. Antisera raised against expressed portions of the gene sequence cross-react with a protein, now called dystrophin, present in normal adult and fetal muscle but absent in DMD patients.

The identification of the dystrophin gene by positional cloning shows how an uncharacterized trait can be analyzed by first finding the gene and then working back to the product, i.e. genetics in reverse. It also shows how conventional genetics, recombinant DNA technology, sequencing and biochemical analysis need to be combined in order to be successful. There are a number of negative features of positional cloning. First, it is an arduous task and the major successes have involved the efforts of very large research groups. Second, mapping the trait of interest (e.g. cystic fibrosis) to a narrow region of the chromosome is difficult if it is not associated with any obvious chromosomal abnormality. For the positional cloning of the cystic fibrosis gene use was made of the phenomenon of linkage disequilibrium that is described later in this chapter in connection with the mapping of polygenic diseases (see p. 102).

The candidate gene approach

In contrast to positional cloning, the candidate gene approach does not require the isolation of new genes but relies on the availability of information regarding function and map position from previously isolated genes. An example of this approach is the first detailed analysis of the hereditary eye disease retinitis pigmentosa (RP). The defects causing RP map to a number of different chromosomal locations but the phenotype in each case is photoreceptor degeneration. In one large pedigree the RP gene was mapped to the long arm of chromosome 3. Coincidentally, the gene for the photoreceptor protein rhodopsin also had been mapped to this region. This suggested that mutations in the rhodopsin gene might account for those cases of RP that map to chromosome 3 and sequencing of the gene from affected individuals identified a single nucleotide change. As expected, those individuals with RP mapping to other chromosomal locations did not have the rhodopsin mutation.

The candidate approach to gene identification and characterization has been greatly facilitated by the large number of genome sequencing projects plus advances in bioinformatics. For example, the human genome has been completely sequenced and many of the genes in the sequence have been identified. A key task now is to establish the function of the genes that have been identified. One way of doing this is to start with the DNA sequence for each gene, convert it into the corresponding protein sequence, and then search the bioinformatics databases for related proteins whose function is known. As this information on gene function accumulates it can be used to analyze gene defects. For example, when a new trait is assigned to a specific map position, the genomic database is interrogated for genes located in the same region. The functions of these genes are compared with the features of the trait to find the most likely candidate gene. The potential gene from individuals exhibiting the trait can be screened by sequencing or hybridization to determine whether it carries sequence abnormalities. The availability of the mouse genome sequence will greatly facilitate these efforts (Box 4.2).

Box 4.2 The mouse genome sequence and its relevance to human disease

The sequence of the mouse genome was published in December 2002 and should facilitate the search for human disease genes and their biochemical analysis. There are a number of reasons why this should be. First and foremost, the mouse genome has a similar number of genes as the human genome, 99% of these genes appear to be identical, and 96% of them are syntenic (in the same gene order). This means that disease genes identified in the mouse can be transposed to the human gene map. Second, the mice strains used in laboratory experiments are inbred whereas humans are extensively outbred. Even when one studies large pedigree families, each chromosome pair has a different lineage whereas in mice the chromosome pairs are identical. Third, experimental crosses can be made between mice with different characteristics and the resultant offspring can be analyzed very soon thereafter. Fourth, mice can be mutated in order to generate animals with particular gene defects whose phenotype then can be investigated. These points are elaborated below.

More than 1000 spontaneously arising and radiation-induced mouse mutants causing hereditable Mendelian phenotypes are known. Largely through positional cloning, the molecular defect is now known for about 200 of these mutants. The availability of an annotated mouse genome sequence should increase efficiency since it will be possible to go from genetic mapping to identification of candidate genes. The mouse genome sequence will be crucial in efforts to exploit the growing repertoire of mutant mice that are being generated by chemical mutagenesis. Ten large scale mutagenesis centers have been established worldwide and these are focusing on dominant and recessive screens for a large number of clinically relevant phenotypes. For each mutant, identification of the molecular defect will require positional cloning.

The availability of more than 50 inbred strains of mice, each with its own phenotype for a variety of continuously variable traits, has provided the ability to map genes involved in complex genetic disorders. A systematic study is underway to define parameters such as body weight, behavioral patterns and disease susceptibility among a standard set of inbred lines. Appropriate crosses between such lines followed by genotyping will enable these complex traits to be mapped. Using the mouse genome sequence as a guide the effort required to positionally clone these traits should be greatly simplified.

In recent years a large number of mutant mice have been generated that have specifically engineered loss- or gain-of-function mutations in specific genes of medical or biologic interest. These so-called "knockout" and "knockin" mutants have been constructed by the application of gene manipulation techniques to embryonic stem cells (see p. 200). The generation of these mutants will be greatly facilitated by the availability of the mouse sequence since experimental design will be simplified. In addition, not all mouse models replicate the human phenotype in the expected way. The availability of the full human and mouse sequences provides an opportunity to anticipate these differences. For example, the mouse mutation might not have been made in the exact counterpart of the human gene orthology. This can occur when there are a number of genes of very similar function. Alternatively, the human genome might contain only a single gene family member whereas the mouse genome may contain multiple family members with overlapping biochemical activities.

Two points of caution are worth noting. First, particular mutations in human genes have been identified as the cause of particular diseases. When the mouse genome sequence was analyzed 160 of these mutations were found to exist in mice that were phenotypically normal. Why this should be is not clear, but understanding the basis for the species difference would be very helpful. Second, the mouse genome has a considerable number of gene family expansions when compared with the human genome. These expansions primarily are in gene families associated with reproduction, immunity, olfaction and response to xenobiotics and are not really surprising when one considers the habitats and lifestyle of mice. Of significance to the pharmaceutical industry is the expansion of the cytochrome P450 gene family since this means that mice and humans could respond differently to drugs.

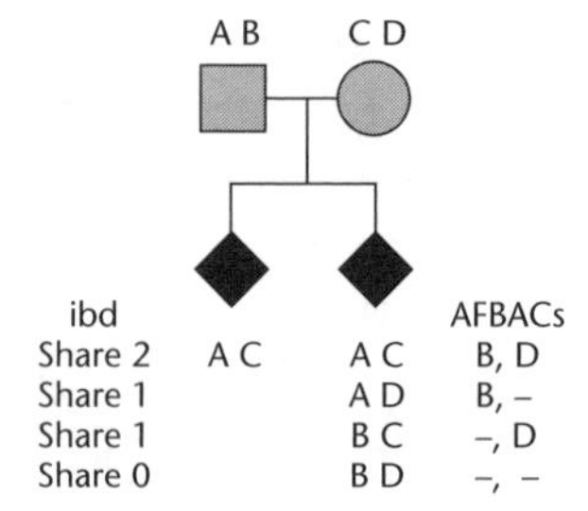

Fig. 4.7 Affected sib pair families. A nuclear family pedigree is shown with the father (gray square) and mother (gray circle) in the first row and the two affected children of either sex (black diamonds) in the second row. Assume for simplicity that we can distinguish all four parental alleles, denoted A, B, C and D in the genetic region under study, with the parental alleles ordered such that A and C are transmitted from the father and mother, respectively, to the first affected child. Four possible configurations among the two offspring with respect to the alleles inherited from the parents are possible: they can share both parental alleles (A and C); they can share an allele from the father (A) but differ in the alleles received from the mother (C and D); they can share an allele from the mother (C) but differ in the alleles received from the father (A and B); or they can share no parental alleles in common. These four configurations are equally likely if there is no influence of the genetic region under consideration in the disease. The parental alleles that are never transmitted to the affected sib pair in each family type are used as a control population in association studies using nuclear family data, the so-called affected family-based control (AFBAC) sample.

Analysis of polygenic disorders

Whereas conventional Mendelian linkage mapping forms a key part of the analysis of single gene disorders, it seldom is useful in the analysis of complex (polygenic) diseases. The involvement of many genes and the strong influence of environmental factors mean that large multigeneration pedigrees are seen only rarely. Consequently, other mapping methods are required and two are in common use. These are **model-free** (or **nonparametric**) linkage analysis and **association** (or **linkage disequilibrium**) mapping.

Model-free linkage analysis

Model-free methods make no assumption about the inheritance pattern, the number of loci involved, or the role of the environment. Rather, they depend solely on the principle that two affected relatives will have disease-predisposing alleles in common. Consequently, analysis is undertaken of families in which both parents and at least two children (sibs) have the disease in question. These are known as nuclear families and the way in which this analysis is undertaken is shown in Fig. 4.7. Suppose that we believe that a certain region of the genome is implicated in a disease state and that we can distinguish the four parental chromosomes

Box 4.3 The linkage between type 1 diabetes and the MHC

There are two major types of diabetes mellitus: juvenile-onset or insulin-dependent (type 1) and adult-onset or insulin-independent (type 2). Type 1 diabetes has a frequency of 0.5% in the Caucasian population and results from an autoimmune destruction of the insulin-producing cells in the pancreas. Genetic factors alone do not cause type 1 diabetes because if one twin of an identical pair develops the disease there is only a 40% chance that the matching twin also will become diabetic. Nevertheless, there is strong evidence for genetic factors and, as noted on page 107, the first study on model-free analysis of a complex disease linked type 1 diabetes with the MHC locus. Individuals heterozygous for HLA-DR3 or HLA-DR4 are particularly susceptible to diabetes. This fits with the concept of type 1 diabetes being an autoimmune disease since DR3 and DR4 are found in a locus known to regulate the immune response.

Further insight into the mechanism responsible for type 1 diabetes has come from a molecular analysis of the HLA-DQ genes. The presence of aspartic acid at position 57 of the DQβ chain is closely associated with resistance to type 1 diabetes, whereas other amino acids at this position confer susceptibility. About 95% of patients with type 1 diabetes are homozygous for DQβ genes that do not encode aspartate at position 57. Since position 57 of the β chain is critical for antigen binding and presentation to T cells, changes in this amino acid could play a role in the autoimmune response that destroys the insulin-producing cells.

(A,B,C,D). If the region under test does not carry a gene predisposing to disease, then the chance of two affected children having two, one or no parental chromosomal regions in common are 25, 50 and 25% respectively. On the other hand, deviation from this Mendelian random expectation indicates that the affected children have chromosome regions that are **identical by descent** (ibd), suggesting the presence of genes predisposing to the disease in question.

Physical markers, particularly microsatellites, are ideal for distinguishing the chromosome regions derived from each parent since they are highly polymorphic and are scattered throughout the genome. In practice, DNA from affected brothers and/or sisters is systematically analyzed using large numbers of physical markers scattered across the genome, i.e. a genome scan. The objective is to find regions that are shared by the two sibs significantly more frequently than expected on a purely random basis. The first complex disease analyzed in this way was type 1 diabetes and linkage to the major histocompatibility complex was discovered (Box 4.3). Since then the method has been extended to other complex diseases and quantitative trait loci controlling adult height.

Linkage disequilibrium mapping

Association or linkage disequilibrium (LD) studies compare marker frequencies in unrelated cases and controls, and test for the co-occurrence of a marker and the disease at the population level. A significant association between a marker and a disease may implicate a candidate gene in the etiology of a disease. Alternatively, an association can be caused by LD of marker allele(s) with the gene predisposing to

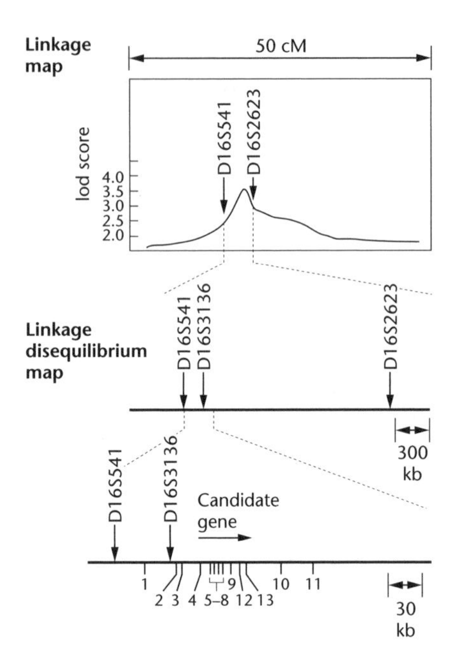

Fig. 4.8 Details of the mapping of a locus on chromosome 16 associated with Crohn's disease. The numbers along the bottom line correspond to the SNPs used in fine mapping. All the SNPs except 10 and 11 showed tight linkage (see text for further details).

disease. LD implies close physical linkage of the marker and the disease gene. As might be expected, LD is not stable over long time periods because of the effects of meiotic recombination. Thus, the extent of LD decreases in proportion to the number of generations since the LD-generating event. Also, the closer the linkage of two markers, the longer LD will persist in the population.

A study of Crohn's disease, a destructive inflammatory disease of the bowel, shows how LD can be used to map disease loci (Fig. 4.8). Conventional linkage analysis had mapped a susceptibility locus to chromosome 16. With the aid of 26 microsatellite markers the locus was mapped to a 5-Mb region between markers D16S541 and D16S2623. LD analysis showed a weak association of Crohn's disease with D16S3136 that lies between the other two markers. A 260-kb region around marker D16S3136 was sequenced but only one characterized gene was identified and this did not appear to be a likely candidate. Sequencing also identified 11 SNPs and three of these showed strong LD with Crohn's disease in 235 affected families indicating that the susceptibility locus was nearby. This turned out to be the *NOD2* locus and it transpired that some of the SNPs used in the study were the causative mutations. It should be noted, though, that the *NOD2* locus confers **susceptibility** to the disease but is not the **cause** of the disease. Other chromosomal loci have been implicated in Crohn's disease but the number of genes involved and the way that they interact to cause disease is not known.

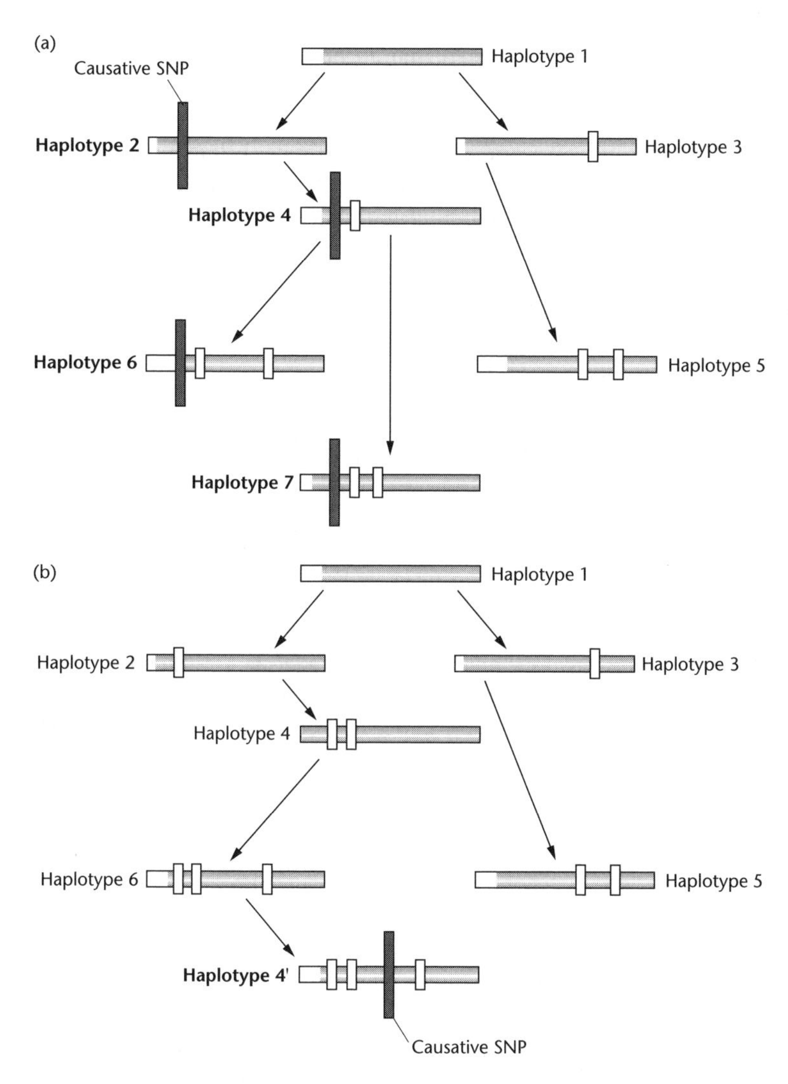

Fig. 4.9 The two ways in which a causative single nucleotide polymorphism (SNP) can become associated with a particular haplotype. In (a) the causative SNP arises early, whereas in (b) it arises late.

Haplotypes

The pattern of SNPs in a stretch of DNA is known as the haplotype. Figure 4.9 shows the evolution of a number of theoretical haplotypes, some of which include an SNP causing a disease. From this figure it is clear that not all SNPs would be predictive of the disease. Also, not all haplotypes are informative and their detection in an LD association study would complicate interpretation of the data. This is exactly what was found in the study on Crohn's disease described above. Nevertheless, when all the SNPs in a particular chromosomal region are analyzed in a large group of unrelated individuals, the results show a picture of discrete haplotype blocks of tens to hundreds of kilobases, each with limited diversity but punctuated by apparent sites of recombination. The existence of haplotype blocks greatly simplifies LD analysis. Rather than using all the SNPs in a region, we can identify exactly which SNPs will be redundant and which will be informative in association studies.

The identification of haplotypes has many implications for medical practice but just two of them will be considered here: HLA haplotypes and responses to drugs.

The major histocompatibility complex

Higher animals, including humans, are able to distinguish between "self" and "nonself" and to mount a reaction against a very broad spectrum of foreign antigens. This reaction is mediated by the immune response. Genetic factors play a key role in the generation of the normal immune response and, as a result of mutation, in aberrant immune reactions including immunodeficiency and autoimmune disease. A large number of genes play a role in the development and functioning of the immune system but only those of the major histocompatibility complex (MHC) are considered here.

The MHC is composed of a large cluster of genes located on the short arm of chromosome 6. On the basis of structural and functional differences these genes are divided into three classes and each class is highly complex and polymorphic. Two of the three classes correspond to the genes for human leukocyte antigens (HLA) that are cell surface proteins. These antigens are very important for the normal functioning of the immune system and were first discovered following attempts to transplant tissue between unrelated individuals. A class I antigen consists of two polypeptide units, a polymorphic peptide encoded by the MHC and an invariant polypeptide encoded by a gene outside the MHC. Class two molecules are heterodimers of α and β subunits, both of which are encoded by the MHC. The class III genes are not HLA genes but include genes for polymorphic serum proteins and membrane receptors.

The HLA system comprises many genes and is highly polymorphic with many antigenic variants having been recognized at the various loci (Table 4.3). Because the HLA alleles are so closely linked they are transmitted together as haplotypes. Each individual has two haplotypes, one on each copy of chromosome 6 and the alleles are co-dominant. Each child receives one haplotype from each parent (Fig. 4.10) and there is a 25% chance that two children with the same parents inherit matching HLA haplotypes. Because the success of tissue transplantation is closely linked to the degree of similarity between HLA haplotypes, the favored donor for bone marrow or organ transplantation is a brother or sister who has an identical HLA haplotype.

Table 4.3 Protein and DNA variation at HLA loci. Because of the redundancy of the genetic code it is possible to have more DNA sequence variants than protein variants.

HLA locus	Antigenic variants (no.)	DNA variants (no.)
HLA-A	25	83
HLA-B	53	186
HLA-C	11	42
HLA-DR (β chain only)	20	221
HLA-DQ (α and β chains)	9	49
HLA-DP (α and β chains)	6	88

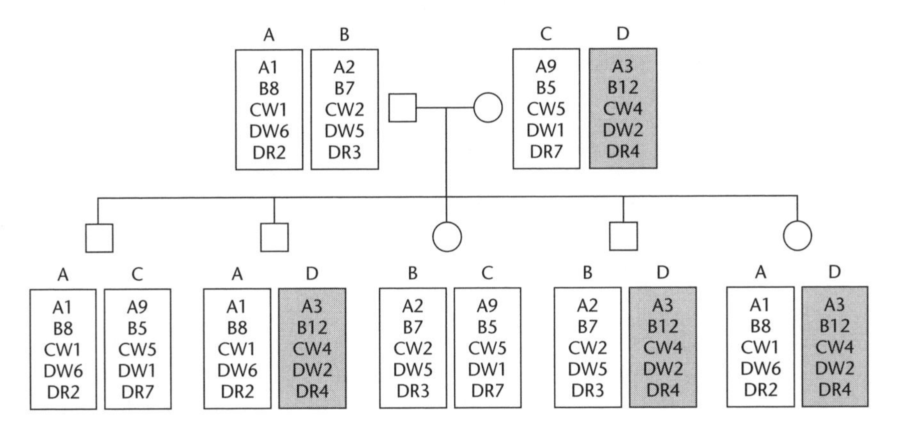

Fig. 4.10 The inheritance of HLA haplotypes. Usually a haplotype is transmitted, as shown in this figure, as a unit. In extrememly rare instances, a parent will transmit a recombinant haplotype to the child.

As more and more information has accumulated about the HLA genes it has become clear that there is an association between specific HLA genes or haplotypes and certain diseases. For example, in one national study, only 9% of the population had the HLA-B27 allele but it was present in 95% of those with the chronic inflammatory disease ankylosing spondylitis. Similarly, 28% of the population carried the HLA-DQ2 allele but it was present in 99% of the population with celiac disease. It is unlikely that HLA genes alone are responsible for specific diseases. Rather, they may contribute to disease predisposition along with other genetic and environmental factors. For example, as noted on page 70, they probably influence the susceptibility of different individuals to particular infectious agents. They also can play a role in complex diseases as exemplified by type 1 diabetes (Box 4.3).

Box 4.4 Social and ethical issues

The ability to screen individuals for mutations that either cause disease or predispose individuals to disease raises a number of social and ethical issues. There are different issues depending on whether the screening is undertaken prenatally or postnatally. For the prenatal detection of inherited defects, use is made of the technique of amniocentesis to withdraw some of the amniotic fluid that surrounds the fetus. This fluid contains some cells (amniocytes) from the developing child and, if desired, these can be cultured *in vitro* to increase their number. These cells then can be probed for particular genetic defects and, if one is found, a decision can be taken to abort the fetus. In many countries there are strict laws against abortion and even in those countries where it is allowed the morality of abortion is strongly debated. Even if the morality of abortion is accepted, where does one draw the line? For example, if the prenatal testing shows that the developing child will suffer from cystic fibrosis or severe combined immunodeficiency then a case can be made for abortion. However, suppose that the fetus is female and tests positive for one of the cancer susceptibility genes *BRCA1* or *BRCA2*. These mutations increase the risk that the carrier will develop cancer but do not guarantee that cancer will occur. Should the fetus be aborted because of this increased risk? Again, suppose that a gene or group of genes can be shown to result in homosexuality (as was suggested a few years ago). Would it be right for heterosexual individuals opposed to homosexuality to abort a fetus at risk of being homosexual?

The principal issue arising from postnatal genetic screening is genetic discrimination, which is defined as the denial of insurance or health care as a result of one's genetic composition. In the USA, Afro-Americans have long been subject to discrimination of employment. Because they have a high incidence of sickle cell anemia, Afro-Americans were required to prove that they were free of the disorder as a condition of employment. Similarly, carriers of the sickle cell mutation (i.e. heterozygotes) were banned from being pilots in the US Air Force on the grounds that they might be unable to cope with reduced oxygen pressures at high altitudes. Such discrimination is in breach of laws against racial prejudice but, more recently, new laws banning genetic discrimination also have covered it.

Prior to the introduction of laws banning genetic discrimination, health insurance companies refused to provide insurance cover for individuals who had subjected themselves to genetic testing and been found to carry a genetic disorder. Indeed, many insurance companies specifically asked applicants if they had had any genetic tests and, if so, what were the results. Failure to provide this information was grounds for refusal of insurance cover. The issues here are the privacy and confidentiality of an individual's genetic information and the economic and legal questions that arise from this information. In this context, it is worth noting that in many countries military personnel are required to provide a DNA sample to facilitate identification of badly mutilated bodies arising from armed conflict or other disasters. These DNA samples could be used for genetic testing without the consent of the donor, although this is prohibited in most countries by laws governing genetic privacy. Similarly, some countries maintain DNA profiles of convicted criminals and these also could be used for genetic testing.

Most genetic testing is undertaken by clinical geneticists working within the normal health care system. However, some private companies market genetic tests directly to consumers and such tests give cause for concern. First, many of these companies do not give information about confidentiality or warn of the risk of genetic discrimination. Second, the customer may misinterpret the information that these tests provide. The *BRCA1* and *BRCA2* genes provide a good example of the problem. These genes encode tumor suppressors that prevent cells from dividing in an uncontrolled way. If the genes are missing or altered then the individual has a 50% higher chance of developing breast cancer and the risk of developing ovarian cancer also increases. If the woman tested is a member of a high-risk family, i.e. those with several cases of both breast and ovarian cancer among first-degree relatives in successive generations, and she has a positive test, the probability that she will develop breast cancer may be as high as 90%. In such cases, prophylactic surgery to remove breasts and ovaries greatly increases the chances of avoiding the disease. For all other individuals who test positive the risk of developing cancer is only slightly increased above normal but there are many cases on record of such individuals voluntary having their breasts removed.

Sociologists studying genetic testing have made some interesting observations. First, individuals tend to be very polarized towards vigilance or avoidance. Vigilant individuals accept the diagnosis, search for as much information about the disease as possible, and often become politically active by advocating more funding be devoted to research and health care. Avoiders use a variety of strategies to play down, or even hide, the presence of the disorder in their family. Second, there is a very marked gender difference: men tend towards avoidance and women to vigilance. Third, within the USA, Afro-Americans and Latinos are much less likely than white and Asian-Americans to voluntarily be screened genetically. This finding may be related to the history of discrimination against individuals with sickle cell disease.

Individual responses to drugs (pharmacogenomics)

There are two fundamental causes of individual responses to drugs: variation in the structure of the target molecule and differences in drug metabolism. For example, the beta$_2$-adrenergic receptor agonists are the most widely used agents in the treatment of asthma and several polymorphisms have been described within target genes. Several studies have shown associations between SNPs in these genes and response to therapy. One study has shown that homozygotes for one allele were up to 5.3 times more likely to respond to albuterol than homozygotes for the alternative allele. Heterozygotes were 2.3 times more likely to find albuterol beneficial.

Many adverse reactions to drugs are due to a failure of metabolism. As an example, between 3 and 10% of the Caucasian population fail to metabolize the adrenergic blocking drug debrisoquine and treatment results in severe hypotension. In Afro-Americans the frequency of this "poor metabolizer" condition is 5% and in Asians it is just 1%. Affected individuals are homozygous for a mutant cytochrome P450 gene (*CYP2D6*) and they also fail to metabolize over 20% of all commonly prescribed drugs, including codeine. The same gene also has alleles that cause an elevated-metabolizer phenotype and this has been correlated with increased susceptibility to cancer.

For many genetic polymorphisms affecting drug efficacy there is no evident phenotype in the absence of a drug challenge. This brings an unwanted element of chance into the selection of appropriate therapies for patients and the selection of patients for clinical trials of new drugs. In both cases there are very significant cost implications. If the genotype of an individual was known in advance then better clinical decisions could be made. Because many of the polymorphisms causing adverse effects are the result of single nucleotide changes, then a SNP profile of an individual could be used to guide therapy or selection for participation in a clinical trial. Given that the human genome is 3×10^9 basepairs in size and that SNPs occur on average every 1000 nucleotides, there are over 3 million SNPs in the human genome. Typing an individual for all of them would be a gargantuan task and one that would need to be repeated many times. However, with the existence of haplotypes the typing response is much smaller – but still daunting! Because of the very high cost (hundreds of millions of

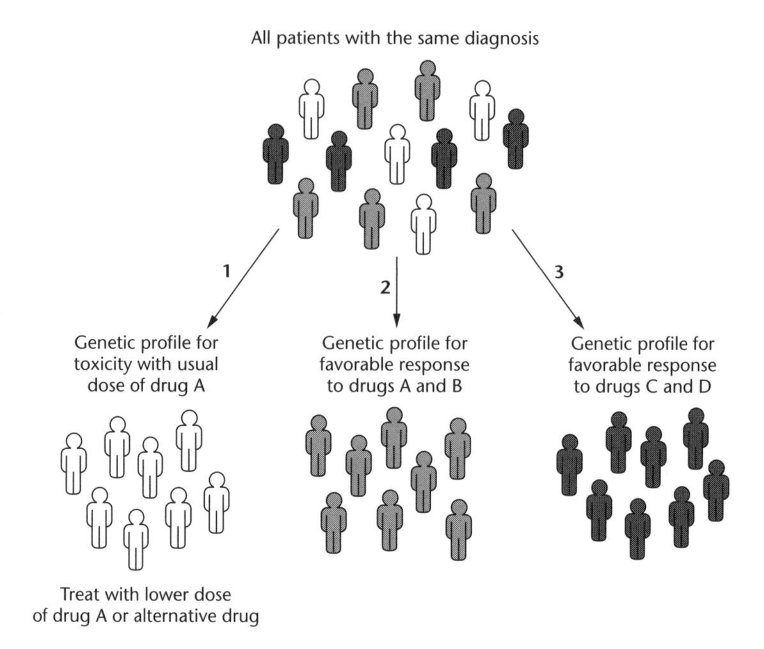

Fig. 4.11 Pharmacogenomics has the potential to subdivide a population of patients with the same empiric diagnosis (e.g. hypertension) into subgroups that have inherited differences in their metabolism of and/or sensitivity to particular drugs. One subset of the population might be at substantially greater risk of serious toxicity (1), whereas other subsets may have receptor polymorphisms or disease pathogenesis polymorphisms that make them more responsive to different treatment options (2 vs. 3).

dollars) of undertaking clinical trials of a new drug, especially if the drug fails at a late stage, pharmaceutical companies are spearheading the development of highly automated high-throughput SNP profiling. Initially this will lead to better classification of the different subtypes of a disease and hence to better diagnoses. This in turn will facilitate selection of the most appropriate therapies. Later, when we know the exact cause of each disease, we should be able to develop drugs that will treat the cause rather than the symptoms as occurs at present. Ultimately, genetic analysis of affected individuals will suggest what drugs **could** be used and pharmacogenomic analysis will determine which drugs **should** be used (Fig. 4.11). This will be the era of **personalized medicine** (Box 4.4).

Further reading

POGM: Chapter 2 covers in detail the principles of nucleic acid blotting and the different blotting techniques (Southern, northern, etc).

POGA: Chapter 6 details how bioinformatics can be used to identify genes from sequence data. Chapter 12 explains many of the specialist terms used in human genetics, including the use of LOD scores, and has a section devoted to quantitative traits.

Alper JS, Ard C, Asch A, *et al.* (eds) (2002) *The Double-Edged Helix: Social Implications of Genetics in a Diverse Society*. Johns Hopkins Press, Baltimore.

Johnson JA, Evans WE (2002) Molecular diagnostics as a predictive tool: genetics of drug efficacy and toxicity. *Trends Mol Med* **8**, 300–305.

Roden DM, George AL (2002) The genetic basis of variability in drug responses. *Nature Reviews Drug Discovery* **1**, 37–44.

Two excellent papers reviewing the consequences of human genetic variation on drug response.

Judson R, Stephens JC, Windemuth A (2000) The predictive power of haplotypes in clinical response. *Pharmacogenomics* **1**, 15–26.

This review stands out as a model of clarity in a large field of papers that are almost impossible for the nonspecialist to understand.

Nussbaum RL, McInnes RR, Willard HF (2001) *Genetics in Medicine*. WB Saunders, Philadelphia.

An excellent text which has very comprehensible sections on most of the topics covered in this chapter.

Peltonen L, McKusick VA (2001) Dissecting human disease in the postgenomic era. *Science* **291**, 1224–1229.

An excellent overview of the way in which genomics and classical genetics are being combined to unravel genetic diseases.

Reich DE, Cargill M, Bolk S *et al.* (2001) Linkage disequilibrium in the human genome. *Nature* **411**, 199–204.

An analysis of linkage disequilibrium in different populations.

Twyman RM, Primrose SB (2003) Techniques patents for SNP genotyping. *Pharmacogenomics* **4**, 67–79.

This paper is a detailed review of all the methods used and proposed for identifying SNPs.

The 26 April 2002 issue of the journal *Science* was devoted to the study of complex diseases.

CHAPTER FIVE

Diagnosis and treatment of cancer

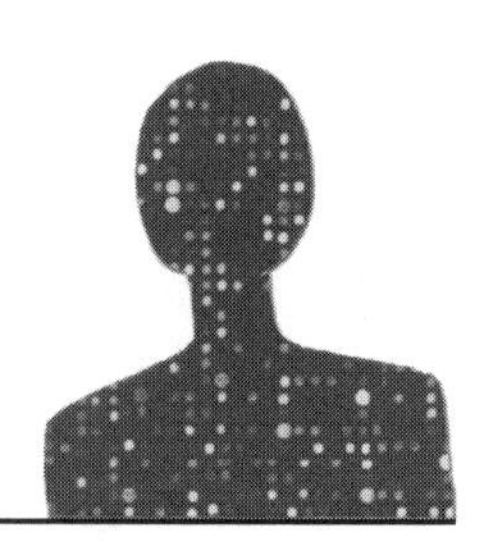

Introduction

Cancer is the general term for any disease that involves the abnormal and uncontrolled proliferation of cells. The new cells, which serve no useful function in the body, are described as **neoplastic**. In many forms of cancer, the neoplastic cells clump together to form a **tumor**. **Benign tumors** contain cells that are similar to normal cells, and because they are enclosed in capsule of fibrous extracellular material, they stay in one place. By contrast, **malignant tumors** contain highly abnormal cells that can break away and spread to different parts of the body, seeding further tumors. This spreading process is called **metastasis** and is difficult to control. It is therefore very important to diagnose and treat cancer as early as possible.

There are over 200 different forms of cancer in humans, and a wide spectrum of different tissues and cell types may be involved (Table 5.1). The best treatment, or combination of treatments, varies according to the specific type of disease. There are three broad classes of cancer, known as carcinomas, sarcomas, and hematologic cancers. **Carcinomas** are tumors that arise from epithelial cells lining internal and external surfaces of the body, e.g. the skin, the stomach lining, the breast ducts, and the lungs. **Sarcomas** arise from cells in the supporting structures of the body, e.g. bone, muscle, and cartilage. Hematologic cancers arise from blood cells or their progenitors; they include **leukemias, lymphomas**, and **myelomas**.

The molecular basis of cancer

Unlike the diseases discussed in the previous two chapters, cancer is neither inherited nor infectious. Cancer is a genetic disease, because it always involves the alteration of DNA, but it is acquired in **somatic cells** during an individual's lifetime and is not passed on to the next generation. Ultimately, the onset of cancer is due to mutations that affect the genes controlling cell growth and proliferation. One mutation in one gene is not enough to cause cancer because humans have evolved complex regulatory networks to control cell division, complete with backups and

Table 5.1 Some of the commonest forms of cancer.

Disease	Comments
Lung cancer	Leading cause of cancer death in both sexes. Strong links to smoking (active and passive) and environmental hazards such as asbestos and arsenic
Breast cancer	Second leading cause of cancer death in women (rare in men). Risk increases with age. Susceptibility runs in families
Prostate cancer	Second leading cause of cancer death in men. Risk increases with age
Colorectal cancer	Third leading cause of cancer death in both sexes. Links to high-fat, low-fiber diets. Susceptibility funs in families
Skin cancer	Occurs as basal (squamous) skin carcinoma or malignant melanoma. Second most common form of cancer in the under thirties. Exposure of skin to chemical mutagens and radiation, including UV light, can contribute
Ovarian cancer	Occurs mainly in women over 40. Linked to high estrogen levels and poor diet
Leukemia	Acute or chronic types. Caused by uncontrolled proliferation of white blood cells. Patients prone to hemorrhaging and persistent infections
Cervical cancer	Occurs mainly in women over 30. Increased risk associated with smoking, early age of first intercourse, large number of sexual partners, and some sexually transmitted diseases (e.g. herpes)
Testicular cancer	Most common tumor in young males

failsafes should particular components fail. Cancer is therefore a multistage disease, with neoplastic cells becoming progressively more abnormal and aggressive as mutations accumulate (Box 5.1). Sporadic cancers occur when such mutations accumulate *de novo* in somatic cells, but a **predisposition** toward cancer may be inherited if one or more mutations is already present at conception. Therefore, while several human cancers have a very strong tendency to run in families, tumors themselves are not passed from parents to offspring. Most cancers are caused by a complex interplay of genetic and environmental contributory factors, which either stimulate the proliferation of cells directly or increase the chance of mutations occurring in relevant genes (Table 5.2).

The genetic processes that underlie cancer are outlined in Fig. 5.1. Two important classes of genes can be identified as contributing directly to the loss of growth control:

- **Proto-oncogenes** encode proteins whose function in the body is to promote cell growth and proliferation. Under normal circumstances, these genes are expressed only in rapidly dividing cells and the proteins need to be activated by external growth-promoting signals. If mutations cause such genes to be overexpressed or their protein products to be overactive, then uncontrolled proliferation can occur. Mutated versions of proto-oncogenes that increase the risk of cancer are simply called **oncogenes**. Only one copy needs to be present because the effects of oncogenes are dominant.
- **Tumor suppressor genes** encode proteins whose function in the body is to repress cell growth and proliferation. These genes may act as breaks on the cell

Box 5.1 Cancer as an evolutionary process

Evolution is driven by competition, with those species best able to survive and reproduce replacing others with lower fitness and fecundity. In unicellular organisms, cells with the fastest division rates are the most successful. In multicellular organisms, competition between cells is repressed for the benefit of the organism as a whole. However, when that repression is lifted – as occurs in cancer – then cells lose their sense of duty to the organism and behave as selfish entities once again. Because species survival is more important than cell survival, the human race has evolved with enough regulatory devices in place to prevent cancer in most people until they have passed reproductive age. But in evolutionary terms there is no selective advantage in protecting the elderly from cancer. This is why cancer is predominantly a disease of aging.

Current research suggests that at least six mutations are required in a single cell to disassemble the regulatory devices and allow uncontrolled proliferation. However, each successive mutation causes a greater degree of deregulation, so most cancers pass through characteristic stages of dysplasia (mild overproliferation and cell disorganization) and benign tumor growth before becoming malignant and metastatic. As in any evolving system, cells with the highest proliferation rates will come to dominate. The progression of cancer therefore involves the stepwise selection of cells that show greater and greater abnormality as they lose growth control mechanisms. The chances of any cell acquiring six independent mutations in relevant genes are very small, even if the normal mutation rate is enhanced by environmental factors. Therefore, an important component of cancer progression is the loss of DNA stability, which occurs due to mutations in genes involved in the regulation of DNA replication and repair (see text). The accumulation of mutations in the progression of colorectal cancer is particularly well understood, as shown in Fig. B5.1.

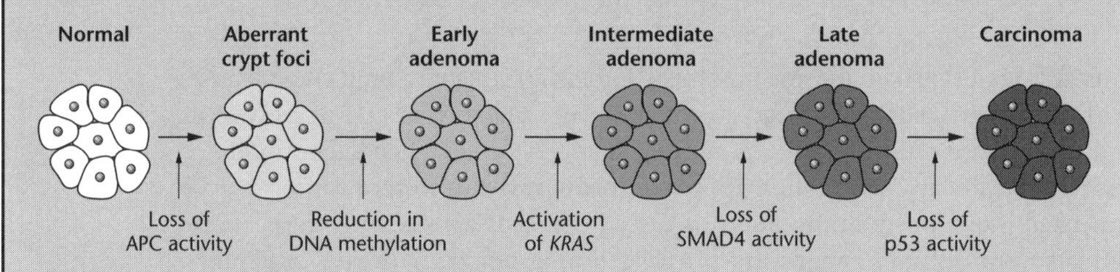

Fig. B5.1 The accumulation of mutations during the progression of colorectal cancer.

division cycle (in contrast to proto-oncogenes, which are accelerators) or they may have alterative functions such as promoting cell differentiation or programmed cell death. If any of these functions fail, there will be an excess of proliferating cells. In most cancers, both copies of the tumor suppressor gene must be inactivated.

The products of proto-oncogenes serve a number of functions in the normal cell, e.g. they may be secreted growth factors, receptors, components of intracellular signaling pathways, transcription factors, or regulatory proteins that promote progression through the cell cycle. The conversion of a proto-oncogene into a cancer-promoting oncogene can occur in four different ways, and some examples are shown in Table 5.3. Note that all four mechanisms involve a **gain of function**. By contrast, since tumor suppressor genes help to prevent cell proliferation, it is their **loss of function** which leads to cancer. This can be caused by whole gene deletions or point mutations. However, it is becoming increasingly evident that in

Table 5.2 Genetic and environmental factors that may contribute to cancer.

Factor	Example
Genetic predisposition	Certain rare mutations make particular forms of cancer so likely that they are often said to **cause** the disease. For example, mutations in the *RB1* gene make retinoblastoma almost inevitable. Mostly, however, predisposition towards cancer is caused by polymorphic variants each with a small cumulative effect. These variants may increase the overall risk of cancer, but other genetic and environmental factors are also important
Gender	Females are more likely to develop breast cancer because breast tissue is naturally more proliferative in females than in males
Environmental hazards	Asbestos, certain industrial chemicals (e.g. benzene), acute high-dose radiation and prolonged low-dose radiation can all cause damage to DNA, increasing the risk of mutations. Smoking is an environmental hazard implicated in 30% of cancer deaths, predominantly lung cancer
Stress	Prolonged stress has been shown to contribute to some forms of cancer although the mechanism is unclear
Food additives	Sodium nitrate can be metabolized into a potent **carcinogen** (a DNA-damaging agent linked to cancer)
Viruses	The risk of cancer is sometimes enhanced by viruses. For example, human papillomavirus and herpesvirus may contribute towards cervical cancer because viral proteins interact with human proteins controlling cell proliferation
Therapy	Estrogen replacement therapy is linked to some cases of breast and ovarian cancer because this hormone stimulates cell proliferation
Lifestyle	Lack of exercise and poor diet have been linked with an increased risk of cancer, although the mechanism is unclear

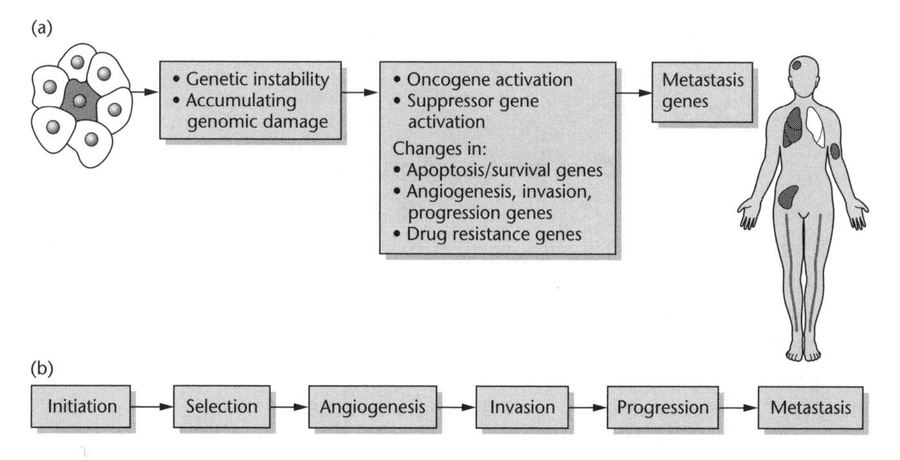

Fig. 5.1 The progression of cancer shown as (a) genetic processes and (b) physical processes.

Table 5.3 Examples of the four mechanisms by which proto-oncogenes are converted into oncogenes.

Mechanism	Gene (product)	Example
Amplification of whole gene, leading to overexpression of normal protein	*ERBB2* (membrane-spanning receptor)	Breast cancer
Point mutation causes overactivity or constitutive activity of protein	*RAS* family (intracellular signalling protein)	Many cancers
Chromosome translocation creates chimeric gene. Fusion protein has enhanced activity	*ABL* (intracellular signaling protein)	*BCR-ABL* fusion gene, chronic myeloid leukemia
Chromosome translocation moves gene from inactive to active chromatin, leading to overexpression of normal protein	*MYC* (transcription factor)	Burkitt's lymphoma

many cancers, the tumor suppressor genes are silenced **epigenetically**, by DNA methylation and the remodeling of chromatin structure. The functions of some key proto-oncogenes and tumor suppressor genes are shown in Box 5.2.

Other classes of gene have indirect effects on cancer progression. The main category is the **tumor-susceptibility genes**, whose function is to maintain genome stability. These genes are often involved in DNA replication and repair, and mutations in them make the genome more prone to further mutations, which increases the frequency of hits on proto-oncogenes and tumor suppressor genes. For example, mutations in the genes *MSH2* and *MLH1*, both of which function in mismatch repair (the correction of mismatched bases after DNA replication), are often seen in hereditary nonpolyposis colon cancer (HPNCC). These tumors show extensive **microsatellite instability** (the tendency for short tandem repeat sequences to expand and contract during successive rounds of DNA replication). Most other types of tumor show **chromosome instability** manifesting as bizarrely abnormal karyotypes with multiple chromosome rearrangements. Further genes are involved in the later stages of cancer. Such genes may, for example, promote angiogenesis (the growth of blood vessels) or the loss of cell adhesion. Drug resistance genes may also be included in this group.

The impact of genomics on cancer research

Unlike other human disease genes, there are several short cuts to the identification of genes involved in cancer. Dominant oncogenes, for example, can be identified by **functional assays** in cell lines that have already lost a number of key growth regulators. Tumor suppressor and tumor susceptibility genes can be identified by **loss of heterozygosity** analysis in the tumors of individuals already lacking one copy of the gene at birth. Essentially, this is a diagnostic approach in which Southern blot hybridization or, more usually, PCR analysis (see Chapter 1) is used to identify segments of DNA that are present in normal cells but absent in tumor

Box 5.2 Molecular control of cell proliferation

Cell proliferation is intimately linked to the control of the cell cycle, which is the series of events occurring between successive cell divisions (Fig. B5.2a). The most important events of the cell cycle, DNA replication (synthesis or **S-phase**) and mitosis (**M-phase**), are separated by gap phases (**G1 and G2**) which allow the cell to grow and accumulate macromolecules. Progression through the cell cycle is controlled by a family of proteins called **cyclins**, which work in concert with **cyclin-dependent kinases (CDKs)**. These cyclin–CDK complexes activate key proteins required at each stage of the cell cycle by phosphorylation, e.g. they activate the components of DNA polymerase at the beginning of S-phase therefore allowing the onset of DNA replication.

The cyclin–CDK complexes are **activated** by external growth factors through signaling pathways, and many of the components of these pathways are therefore **proto-oncogenes** since their increased activity **promotes** cell division (Fig. B5.2b). In order to maintain genome stability, the cyclin–CDK complexes must also be **inhibited** by internal signals that prevent mitosis from going ahead if replication is incomplete, and prevent DNA replication commencing unless cell division is complete. An important component of the regulatory network is the prevention of cell division if there is unrepaired DNA damage: the cell cycle must either arrest until the damage is repaired or, it if is unrepairable, the cell is programed to die by a process called apoptosis. **Tumor suppressor genes** control these processes by acting at so-called checkpoints that monitor progress at certain stages of the cell cycle. In human cells the most important checkpoint occurs at the **G1/S transition** (the onset of DNA replication). Several tumor suppressor gene products act at this point, including the **retinoblastoma protein** pRB and the product of the *TP53* gene, which is simply called **p53**. The loss of p53 function is probably the most important change in the majority of cancers, since this protein plays critical roles in arresting the cell cycle in response to DNA damage and promoting apoptosis. For this reason, p53 has been dubbed the guardian of the genome.

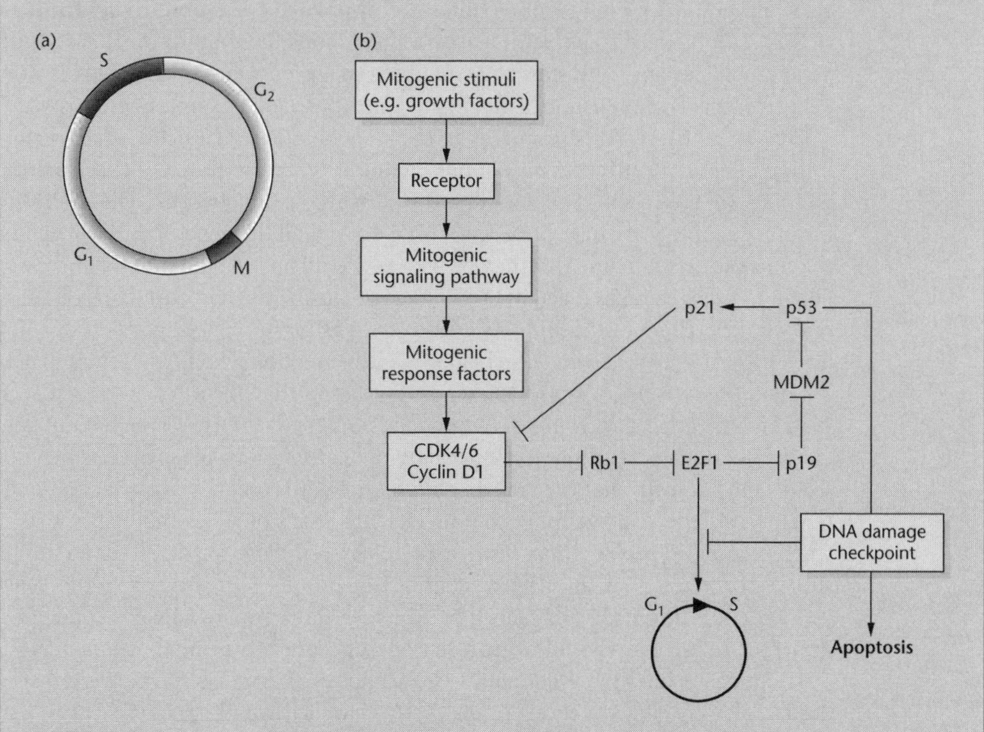

Fig. B5.2 (a) The cell cycle. **(b)** Proteins involved in the regulation of the cell cycle.

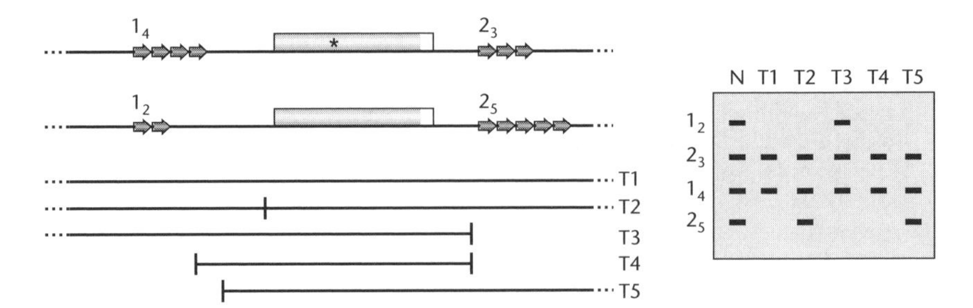

Fig. 5.2 Loss of heterozygosity (LOH) can reveal the presence of tumor suppressor genes. In this example, a candidate tumor suppressor gene is flanked by two microsatellite markers, 1 and 2. The individual in question has inherited one nonfunctional allele (*) and one functional allele of the tumor suppressor gene. The analysis of tumor tissue shows that the microsatellite markers associated with the mutant allele are retained while those associated with the functional allele are eliminated. This suggests the functional allele has been replaced by homologous recombination or deleted as part of a larger deletion also involving flanking markers. While LOH has been used to map several tumor suppressor genes, many tumor samples must be characterized to determine which deletions are causative and which merely represent random genomic instability.

cells (Fig. 5.2). A more recent technique, **comparative genome hybridization (CGH)**, can be used to pinpoint areas of the genome containing cancer-causing gene amplifications. The standard approach using chromosome spreads is now being replaced by much more convenient assays based on genomic microarrays.

Genomics offers further resources for the identification and functional characterization of cancer genes. Perhaps more importantly, it also provides experimental methods to link them into functional pathways and networks. The complete human genome sequence contains an inventory of all the genes in any particular chromosome region, and this may be used to identify candidate tumor suppressor genes or oncogenes identified in loss of heterozygosity and comparative hybridization studies. Homology searching using the 150 or so cancer genes that have already been identified could also help to identify functional homologs, though preliminary analysis using this strategy has thus far failed to identify new candidates. In a converse approach, comparative genomics has been used to screen out unlikely candidates. Cancer is a disease of animals with relatively long life spans and the protective mechanisms that prevent cancer are therefore recent evolutionary adaptations. While the yeast genome contains its fare share of genes that control DNA stability and cell proliferation, it appears to lack orthologs for the 30 or so tumor suppressor genes that have been identified in humans. In this sense, the yeast genome could be used to subtract all the unlikely tumor suppressor gene candidates from the human genome, allowing researchers to concentrate on the remainder.

The tools of functional genomics (see Chapter 2) are also increasingly being used to identify cancer genes. For example, sequence sampling, microarrays, and proteomics are being used to identify genes and proteins that are upregulated or downregulated in cancer. These data are assimilated at a central resource known as the Cancer Genome Anatomy Project, which aims to assemble a comprehensive

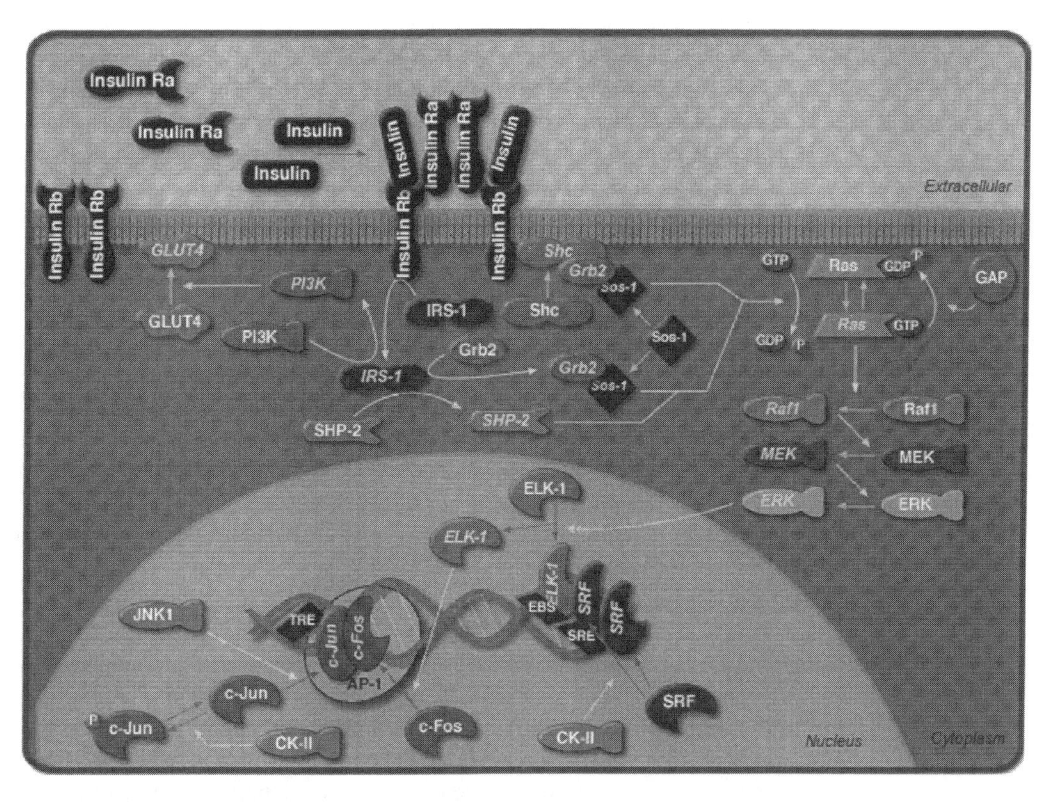

Fig. 5.3 The insulin signaling pathway as represented on the CGAP website. This is one of many cancer-related pathways on the site, together with genes, information about chromosomes and tissues, and expression data. (http://cgap.nci.nih.gov/Pathways/BioCarta/insulinPathway)

molecular resume of every type of cancer (http://www.ncbi.nlm.nih.gov/CGAP/) (Fig. 5.3). Another functional genomics approach with an important role in cancer research is the use of the yeast two-hybrid system and other technologies to map protein interactions. For example, one of the earliest large scale two-hybrid screens involved the use of *Drosophila* cyclin-dependent kinases as baits in order to identify novel cell cycle regulators. And mass spectrometry has been used to identify novel components of the epidermal growth factor signaling pathway, whose overactivity is often implicated in breast cancer. The analysis of protein interactions is particularly useful for the development of anticancer drugs that block cell proliferation by competing for interacting protein surfaces.

New methods for the diagnosis of cancer

The diagnosis of cancer is usually made after physical examination, often followed by the analysis of samples (blood, urine) or tissue biopsies for evidence of neoplastic cells or their products. More sophisticated methods are used to confirm the presence of deep tumors, including for example ultrasound, barium radiography, or a computerized axial tomography (CAT) scan. The impact of biotechnology and genomics on cancer diagnosis is demonstrated by the development of new immunologic techniques for the detection of specific tumors and the use of transcript analysis and proteomics for disease classification and staging.

The development of antibodies as therapeutic agents is discussed in Chapter 6 so we will not go into details here. Suffice it to say that antibodies are capable of binding to target antigens with great specificity, so antibodies recognizing tumor-specific antigens can help to positively identify tumor cells in biopsies and other samples. Furthermore antibodies also recognize tumor cells *in vivo*, and therefore accumulate at the site of a tumor following injection into the bloodstream of cancer patients. The use of antibodies conjugated to radioactive labels is discussed below in the context of radiotherapy, but this approach can also be used in radiography procedures to help locate tumors. Examples of diagnostic antibodies under development include those recognizing carcinoembryonic antigen, which is expressed on the surface of many carcinomas, and human chorionic gonadotropin, which is also synthesized by a number of tumors.

Sometimes it is not possible to use antibodies to identify tumors by the marker proteins they express. Indeed, there is often difficulty in distinguishing between certain forms of cancer using any of the traditional cytological, cytogenetic and biochemical assays. This is the case for **acute myeloid leukemia (AML)** and **acute lymphoblastoid leukemia (ALL)**, where diagnosis is generally based on a battery of tests, none of which is 100% reliable. Importantly, the diseases respond to different drug therapies: daunorubicin and cytarabine work best for AML while ALL patients respond better to vincristine and methotrexate. Therefore a correct diagnosis is essential for a favorable clinical outcome.

Recently, it has been shown that accurate diagnosis can be achieved using DNA microarrays to look at global gene expression profiles (Fig. 5.4). Whereas traditional tests look for single markers, the microarray method involves the analysis of approximately 50 gene expression profiles, which provides a much greater degree of discrimination. Data clustering algorithms are used to process the profiles and place samples into the correct category, a process known as **class prediction**. In one study, 36 out of 38 patients were classified correctly as AML or ALL using this single test. Similar assays are now being developed for a range of solid tumors.

As well as classifying disease subtypes that are already known to exist, the same method can be used to identify new disease subtypes. For example, only 40% of patients with non-Hodgkin's lymphoma respond to current therapy. Until recently, it was impossible to tell which patients would respond until treatment was already underway. However, DNA microarray analysis revealed two, previously unrecognized subtypes of the disease that matched the different clinical outcomes and drug responses. In this case, the data clustering algorithms have no predefined disease categories and the analysis method, known as **class discovery**, is described as **unsupervised**.

While certain cancers can be diagnosed by changes in mRNA levels, it is often more useful to look at proteins since these are the actual functional molecules of the

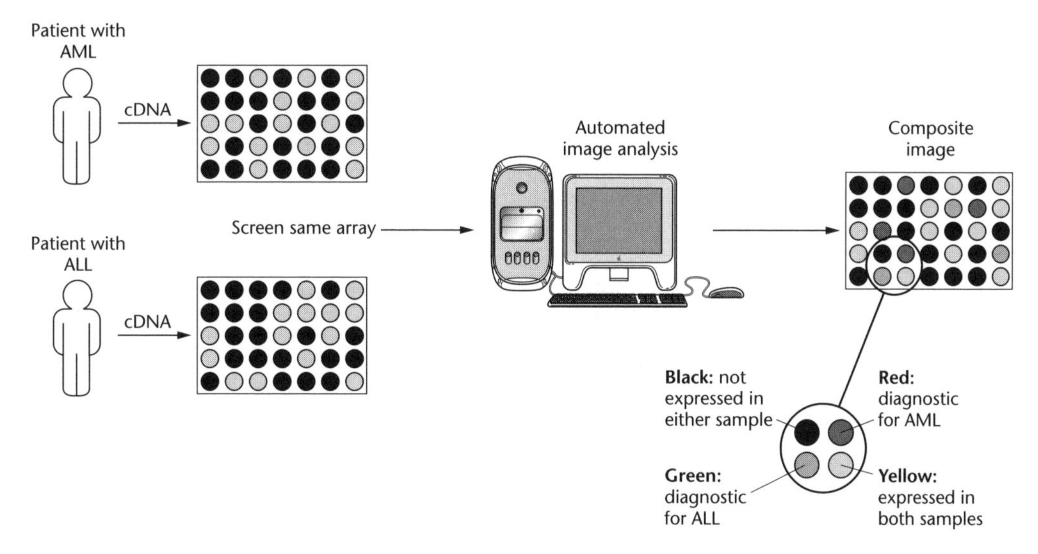

Fig. 5.4 Global expression profiles for cancer diagnosis. The comparative analysis of patients with acute myeloid leukemia (AML) and acute lymphoblastoid leukemia (ALL) reveals genes whose activity is characteristic of one disease or the other. While individual genes are not discriminatory, global expression profiles involving 50–100 genes allow disease diagnosis with increased confidence.

cell. Protein levels may change while the corresponding transcript remains at the same abundance if disease-specific changes occur post-transcriptionally. The core proteomics technology platform – two-dimensional electrophoresis followed by mass spectrometry – can be used to compare normal and disease samples and identify proteins that are suitable disease markers (Fig. 5.5). For example, proteins have been identified that can be used to diagnose breast cancer, colon cancer, and bladder cancer, and in the latter case different proteins have been found that are characteristic of different stages of the disease. This allows accurate staging and appropriate treatment as the disease progresses from the early transitional epithelium to the more aggressive squamous cell carcinoma. Two further examples are discussed below to demonstrate how proteomics is a useful method of diagnosis in situations where transcript analysis would be unsuitable:

• A protein called **stathmin** has been identified that is found at unusually high levels in cases of childhood leukemia. However, only the **phosphorylated** form of stathmin is indicative of the disease. The nonphosphorylated form is found at the same abundance in both healthy and leukemic children. Many proteins are phosphorylated as a regulatory mechanism, and in the case of stathmin (an intracellular signaling protein that relays growth signals) phosphorylation is required for the transduction of the growth signal. While the relative abundance of phosphorylated and nonphosphorylated stathmin can be determined using proteomic methods, both forms of the protein are encoded by the same mRNA so transcript analysis would be uninformative.

• A protein called **psoriasin** has been identified as an early marker of bladder cancer. Psoriasin can be detected in the urine, which like most body fluids is devoid

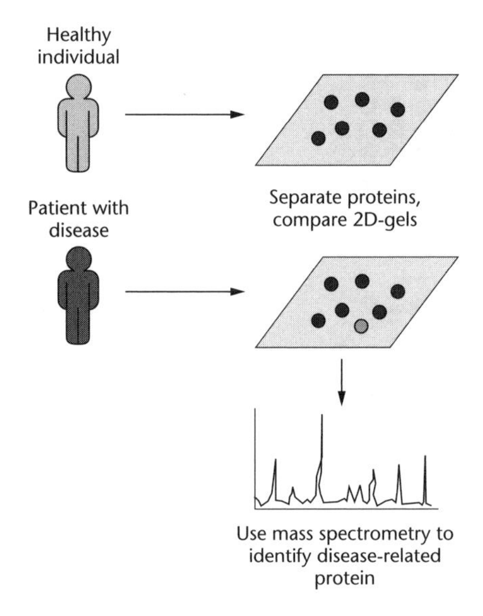

Fig. 5.5 Disease profiling with proteomics. The comparison of healthy and disease protein expression profiles by two-dimensional gel electrophoresis (as shown here) or liquid chromatography–mass spectrometry methods allows the isolation and characterization of proteins that are present or absent in the disease state. Such proteins represent potential disease markers or therapeutic targets.

of mRNA. In this case, proteomics is the best analysis method not because mRNA levels are uninformative, but because there is no mRNA to measure.

New approaches to cancer therapy

There are four basic approaches for the treatment of cancer: surgery, radiotherapy, chemotherapy, and biotherapy. Surgery is still the primary method of treatment in about 30% of cancers, especially tumors caught at an early stage. Surgery may be performed with the aim of curing cancer, but it can also be used for diagnosis or staging (tissue biopsy) and for follow-up care (so called second-look operations, to check the success of initial surgery or alternative treatment). Where the cancer has advanced too far to be cured, surgery is used as a palliative treatment (e.g. decreasing tumor mass to relieve symptoms). The three other forms of treatment have each benefited in different ways from recombinant DNA technology and genomics.

Radiotherapy

Radiotherapy is the treatment of cancer with X-rays or other forms of radiation, applied either from outside the body (teletherapy) or by internal isotopes (brachytherapy). The radiation is thought to work by damaging the DNA of tumor cells. The usefulness of radiotherapy depends on the radiation sensitivity of the tumor as balanced against the sensitivity of surrounding tissues. Sensitivity is partly determined by how much oxygen is available in the tissue, since radiation also generates free radicals that cause secondary damage to DNA. Tumors that are highly sensitive to radiation are easily destroyed with quite low doses, while those with lower sensitivity need higher doses and the impact on surrounding tissues has to be evaluated carefully, particularly if the tumor is on a sensitive organ. Radiotherapy can be used to cure many localized tumors, including those (such as breast tumors) that are operable and those (such as many tumors of the nervous system) that are not. It is also used following surgery to kill residual cancer cells. Radiotherapy has a number of acute side effects such as hair loss and bone marrow depletion, which reflect the effects of radiation on cells of the body that normally divide rapidly. Longer lasting side effects also occur when the radiation kills tissues that renew more slowly.

As discussed above, antibodies can be used to deliver radioisotopes to tumors for diagnostic imaging. However, the use of stronger isotopes such as yttrium-90 (^{90}Y) and iodine-131 (^{131}I) can kill tumor cells very efficiently. Antibodies recognizing tumor-specific antigens help to concentrate the isotope at the site of the tumor, maximizing the killing effect on tumor cells while minimizing the impact on surrounding tissue. This approach is known as **radioimmunotherapy (RIT)** and has been very successful for the treatment of hematologic cancers and in a limited sense for the treatment of certain solid tumors.

A disadvantage of conventional RIT is that the effects of circulating radioactive antibodies often lead to bone marrow and stem cell depletion. Several new **pre-targeting strategies** have been developed in an effort to overcome this problem. For example, a three-step strategy has been used to treat ovarian cancer (Fig. 5.6). In the first step, a biotin-conjugated antibody that recognizes the tumor antigen is introduced into the body. A second component is then introduced that clears circulating antibodies from the bloodstream. Finally, a strepatavidin radioisotope is introduced, delivering high-dose radioactivity directly to the tumor with minimal disturbance elsewhere. This strategy exploits the high-affinity interaction between biotin and streptavidin to achieve effective tumor targeting.

Chemotherapy

Chemotherapy is the use of drugs to treat cancer, particularly "small molecule" drugs as opposed to proteins and nucleic acids. The ideal cancer drug would kill cancer cells selectively while leaving normal cells unharmed. Most of the current anticancer drugs lie a long way from this ideal because they affect all proliferating cells, therefore causing severe side effects. These chemotherapeutic agents are either DNA-damaging chemicals (e.g. alkylating agents such as chlorambucil), antimetabolites (molecules such as methotrexate that interfere with DNA synthesis), or chemicals that interfere with the cell cycle (e.g. paclitaxel, vincristine).

Molecular biology and genomics have a lot to offer in the field of anticancer drug development because they provide the basis for more targeted therapies. The

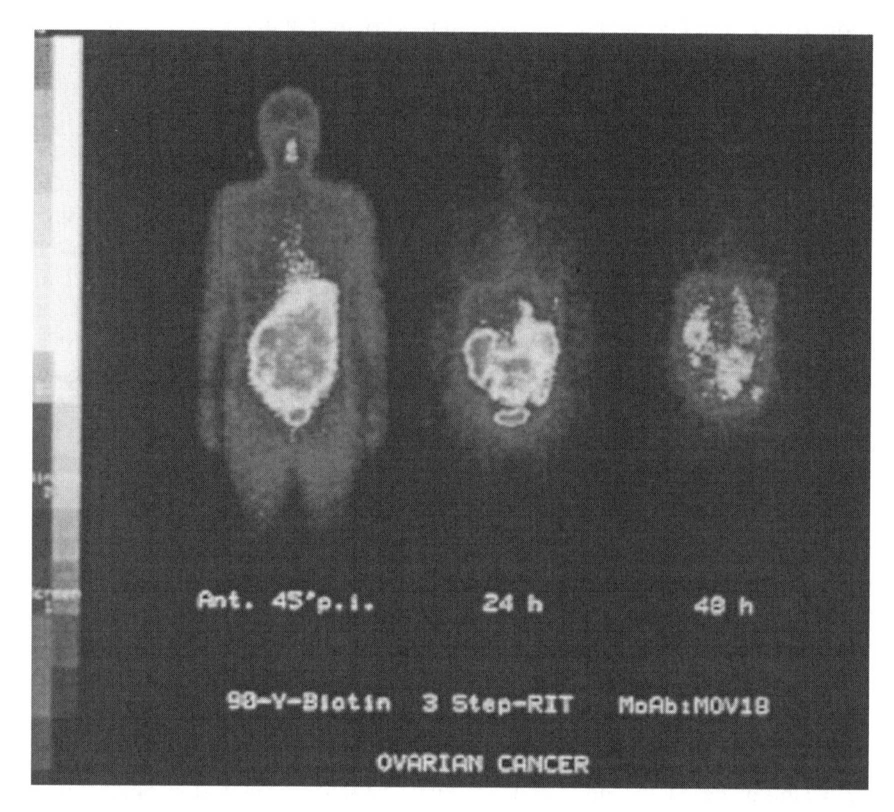

Fig. 5.6 A three-step radioimmunotherapy procedure used to treat a patient with ovarian cancer. The tumor was targeted with a biotinylated monoclonal antibody and after clearing unbound conjugates from the body, a streptavidin–radioisotope complex was injected.

cloning and characterization of oncogenes and tumor suppressor genes generates new targets for drug development, and the selectivity of drugs can be enhanced by studying the structure and interactions of the corresponding proteins. Notably, tumor-specific markers identified by transcript profiling, microarray hybridization and proteomics (e.g. stathmin, see above) represent useful potential targets. Such targets can be validated by gathering information on their expression and biochemical function, and by determining the frequency with which they are deregulated in different types of tumor. If such targets are not suitable (e.g. if the gene shows extensive polymorphism) then proteins that interact with them in complexes or pathways may make useful alternatives. Such proteins can be identified through two-hybrid screens and similar approaches.

Screening programs can then be carried out to identify interacting lead compounds whose therapeutic activity can be optimized. Both large scale screening of chemical libraries and rational drug design can be combined to identify the most suitable candidate drugs, and some examples of emerging or recently approved target-specific compounds are listed in Table 5.4. The first targeted small-molecule

Table 5.4 Some emerging "small molecule" chemotherapeutic agents which target specific proteins that are deregulated in cancer.

Drugs	Target	Function
Iressa (ZD1839)	EGF receptor	Growth factor receptor, overactivity implicated in several cancers
Farnesyl transferase inhibitors (inhibits membrane localization)	Ras family	Relay in growth factor signaling pathway, overactivity a hallmark of many cancers
Wortmanin	PI3 kinase	Regulator of second messengers, required for proliferation, survival, and angiogenesis
Flavopiridol, indirubin	Cyclin-dependent kinase	Cell cycle control
Pithithrin-α (activates p53-dependent genes)	p53	Master tumor suppressor known as the Gatekeeper
PS-341	26S proteosome	Protein degradation, eliminates tumor-suppressor proteins
17AAG	Hsp90	Molecular chaperone, required for the folding and stability of many oncoproteins
Gleevec (ST-1571)	BCR-ABL	Tumor-specific fusion protein, intracellular protein kinase

drug to receive regulatory approval was Gleevec, which binds to the ABL-BCR fusion product characteristic of chronic myeloid leukemia. Model organisms such as *Drosophila* and *Caenorhabditis* have been enlisted in these screening programs because growth factor signaling pathways are highly conserved between humans and these simpler animals. Even yeast cells can be useful for the testing of some drugs, since many components of the cell cycle regulatory network and the basic control of gene expression are conserved between humans and yeast.

Recombinant DNA technology and genomics are also helping to streamline the synthesis of anticancer drugs. For example, some current anticancer drugs are derived from plants, including the mitotic inhibitors paclitaxel, vinblastine, and vincristine. Unfortunately, these molecules are produced in minute quantities which makes them very expensive to extract and limits their use. Paclitaxel is now more widely available because it is produced in plant cell suspension cultures. However, due to the complexities of the metabolic pathways leading to vinblastine and vincristine, a similar production system for these drugs has yet to emerge.

These problems are being addressed by **metabolic engineering** in plants, which is genetic modification to alter the flux along particular metabolic pathways. In the case of vinblastine and vincristine, the plant in question is the Madagascar periwinkle, *Catharanthus roseus*. Both drugs are terpene indole alkaloids, the end products of a long and complex metabolic pathway involving at least 20 enzymatic steps in several different cell compartments. Metabolic engineering can be used to modify certain stages of the pathway, e.g. to remove rate-limiting steps or inhibitors and thus increase the output of desired molecules. For example, a critical early step in the pathway is catalyzed by the rate-limiting enzyme tryptophan decarboxylase,

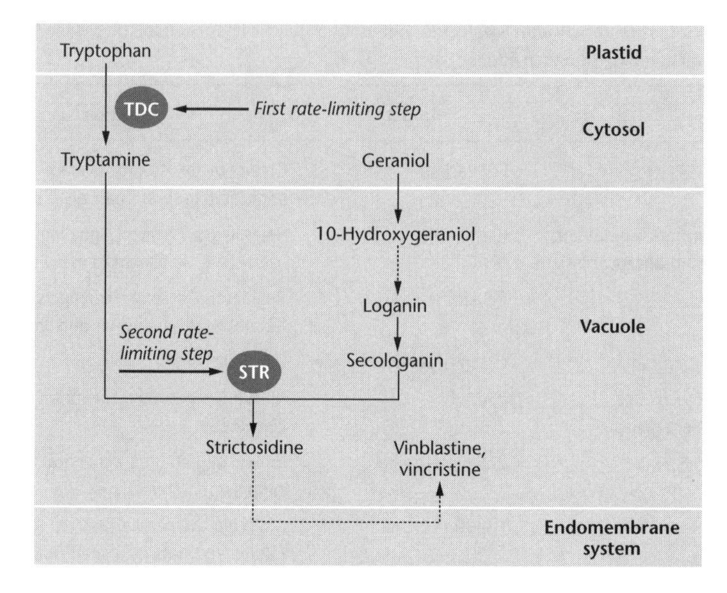

Fig. 5.7 A simplified representation of the metabolic pathway in *Catharanthus roseus* that leads to the production of the anticancer drugs vinblastine and vincristine. Note the multiple cell compartments that are involved and the intersection of two separate pathways when tryptamine and secologanin are converted into strictosidine. These features, as well as the complex regulation at the gene expression and protein activity levels, make the reproduction of the pathway in culture systems very difficult. The enzymes TDC (tryptophan decarboxylase) and STR (strictosidine synthase) are shown. Removing the first bottleneck by increasing the availability of TDC has simply identified the next step in the pathway at STR as another bottleneck. One key to this problem may be the identification and overexpression of transcription factors that regulate multiple genes in the pathway.

so the *tdc* gene has been overexpressed in an attempt to increase the availability of the enzyme and hence increase the output of alkaloids. While this strategy resulted in the accumulation of the immediate reaction product tryptamine, no increase in alkaloid production was evident because the next step in the pathway became limiting (Fig. 5.7).

Genomics is being used to overcome these limitations by helping to characterize metabolic pathways and regulatory networks more thoroughly. Also, high-throughput functional analysis techniques can be used to identify master regulatory genes. For example, the use of an **activation trap construct**, which integrates randomly into the genome and activates adjacent genes (Fig. 5.8), has resulted in the identification of a key transcription factor that co-ordinately regulates several steps in the alkaloid biosynthetic pathway. Another transcription factor with similar properties has been found using a variation of the yeast two-hybrid system (see p. 52) to identify proteins binding to specific DNA sequences. An alternative approach is the transfer of complete metabolic pathways to new hosts. For example, it has been shown recently that vitamin A can be synthesized in rice grains by introducing all the components of the metabolic pathway from

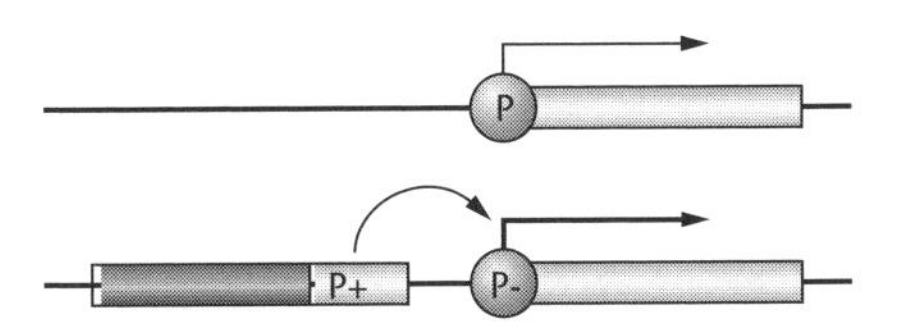

Fig. 5.8 Activation trapping to identify gain of function phenotypes, such as increased alkaloid production in plants. The top panel shows an endogenous plant gene under the control of a weak or restricted promoter. The integration of an activation trap construct, which contains a strong, outward-facing promoter, enhances the activity of the adjacent gene.

other sources. It may be possible in the future to produce drugs such as vinblastine and vincristine in the same way.

Biotherapy

Biotherapy is the most recent development in cancer therapy and involves the use of biologic agents such as proteins (including antibodies), peptides, nucleic acids, viruses, and whole cells. Although the distinction between different types of biotherapy is blurred, we concentrate here on the use of proteins as therapeutic agents and defer the coverage of nucleic acid therapies, including gene therapy for cancer, until Chapter 8. There are currently very few approved biotherapy agents in use but many innovative approaches have been described in the scientific literature and a number of these are undergoing clinical trials. Conventional biotherapy treatments include the use of interferons or interleukins to stimulate the immune system. For example, alfa-interferon is a self-administered protein which enhances the antigenicity of tumors by upregulating genes of the major histocompatibility complex and increasing the activity of cytotoxic T lymphocytes. By contrast, interleukin 2 (IL-2) is administered in a hospital setting. Its main antitumor activity is to stimulate the production of lymphokine-activated killer cells, which infiltrate tumors and release toxins that kill them. Alternatively, IL-2 can be used to stimulate the activity of a patient's cultured T lymphocytes, which are then reintroduced. This strategy is known as **adoptive biotherapy**.

About 20% of pharmaceuticals in current development are antibodies and their derivatives, and many of these are designed for the treatment of cancer. Monoclonal antibodies that recognize tumor antigens trigger the complement cascade and other cytotoxic effector functions, so helping to destroy cancer cells and remove them from the body. However, in some cases there is evidence that antibodies mediate their effects not only by targeting tumor cells for elimination, but also by regulating their growth. For example, anti-idiotype antibodies that recognize the receptors displayed by malignant B lymphocytes have a direct antiproliferative effect which can be used to treat B-cell lymphoma. Herceptin was one of the first therapeutic monoclonal antibodies to receive regulatory approval, and recognizes the ErbB2 receptor tyrosine kinase that is overexpressed in some forms of breast cancer. This antibody is thought to mediate its effects by increasing the rate at which the receptor is internalized and degraded; it shows very encouraging

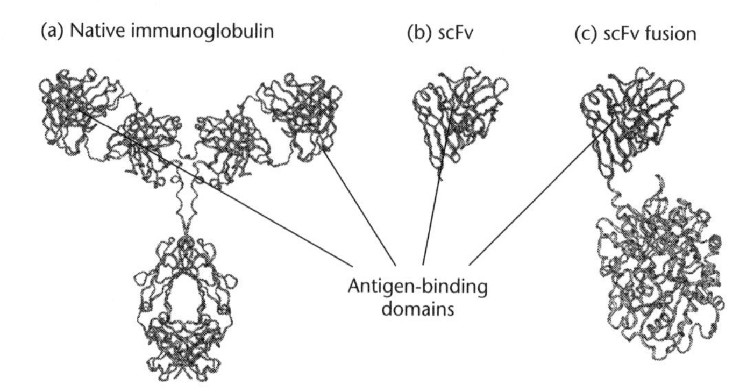

Fig. 5.9 Antibody derivatives used to treat cancer. (a) The structure of a native immunoglobulin, with the variable regions (antigen-binding domains) shown. (b) A single-chain Fv fragment, which comprises the variable regions of the immunoglobulin heavy and light chains joined by a flexible peptide arm. Immunofusion proteins, such as immunotoxins and immunocytokines, are generated by expressing scFv fragments as fusion proteins. (c) The structure of the scFv–interleukin 2 fusion. Other immunodrugs are generated by conjugating antibiotics and other small molecules to scFv fragments or full size antibodies.

results in the treatment of metastatic breast cancer when used in combination with chemotherapy.

More sophisticated antibody derivatives include **immunotoxins**, in which the antibody is conjugated to a potent toxin such as ricin, and **immunodrugs**, where the antibody carries a small molecule such as an antibiotic (Fig. 5.9). Both strategies have been successful for the treatment of different forms of cancer in clinical trials. For example, an anti-CD25 antibody conjugated to ricin has been tested in patients with chemotherapy-resistant forms of leukemia and lymphoma. A variant approach is the expression of an antibody fusion with RNase, although this has yet to be evaluated in the clinic. Immunotoxins appear to work better for hematologic cancers than for solid tumors because cells in the former are not joined by tight junctions that conspire to inhibit penetration. An immunoconjugate called BR96-DOX, which contains the antibiotic doxorubicin, has undergone phase I and II clinical trials for the treatment of breast cancer.

Another recent development is the **immunocytokine**, where a recombinant antibody is expressed as a fusion with a cytokine such as IL-2. The advantage of this approach is that the cytokine, which maintains independent activity, is concentrated at the tumor. Preclinical trials with a murine neuroblastoma model have shown that sufficient immune cells are activated by the attached IL-2 to eliminate established metastases. A final category of immunoconjugates that may be valuable as anticancer agents – **abzymes** – are antibodies conjugated to enzymes. The use envisaged for such molecules is **antibody-directed enzyme/prodrug therapy (ADEPT)**, where the abzyme converts a circulating prodrug (inactive, nontoxic) into a toxic antitumor drug. For example, antibodies specific for carcinoembryonic antigen and conjugated to β-glucuronidase have been used to convert the inactive prodrug glucuronyldoxorubicin into doxorubicin at the tumor site in preclinical trials.

New therapeutic targets

Unraveling the molecular pathways that lead to cancer occasionally results in the revelation of some new and unexpected participants, which represent novel targets for drug development. One example is **telomerase**, the enzyme responsible for adding protective structures called **telomeres** to chromosome ends. Telomerase is present in the cells of the embryo, but its activity declines in most of our cells so our chromosomes shorten by about 100 bp at each round of replication. Cultured primary cells also lack telomerase, and after many rounds of division reach a point of **senescence** where they stop dividing to preserve chromosome integrity. Fibroblasts deficient for p53 and the retinoblastoma protein continue beyond senescence towards **crisis**, where most of the cells die. However, a small number of cells survive and appear to be immortal. These cells have generally reacquired telomerase activity. Similarly, telomerase is active in the majority of tumors *in vivo*.

Another novel group of targets is the enzymes that control DNA methylation and chromatin structure. Epigenetic modifications are observed in many cancers and a growing list of genes is known to be silenced in this manner. These genes include not only tumor suppressor genes, but also those involved in apoptosis, DNA repair, metabolism of potential carcinogens, and metastasis. Drugs that selectively interfere with DNA methylation (e.g. 5-azacytidine) or histone acetylation/deacetylation (e.g. phenylbutyrate) are also undergoing clinical trials.

Finally, angiogenesis (the development of blood vessels in tumor tissue) is a central process in tumor progression. Once tumors grow to a certain size, they must gain an independent blood supply in order to maintain growth. HIF-1α (hypoxia inducible factor) is a key target in this pathway because it is induced by low oxygen conditions and activates 40 or more genes required for blood vessel growth and diversification. Strategies to inhibit HIF-1α and some of its upstream activations (including PI3K, see Table 5.4) can therefore be used to limit tumor growth. Antigens present on developing blood vessels may also represent appropriate targets for immunotherapy, since this approach avoids the requirement for tumor penetration. Antibodies have been raised against a number of vascular targets, including vascular endothelial growth factor, angiogenin, and endoglin. High-throughput expression analysis and proteomics also have a useful role to play as they could help to identify targets expressed on tumor vasculature but not on normal blood vessels.

Further reading

POGM: Chapter 6 describes some of the cloning strategies used to identify proto-oncogenes and their cancer-causing oncogene counterparts.

POGA: Chapter 7 discusses comparative genomics and the conservation of genes and whole pathways between organisms. Chapter 9 goes into more detail about the use of transcript analysis and proteomics to investigate cancer, while Chapter 12 discusses interaction screening.

Boyer TG, Chen P-L, Lee W-H (2001) Genome mining for human cancer genes: wherefore art thou? *Trends in Molecular Medicine* **7**, 187–189.

Futreal PA, Kasprzyk A, Birney E *et al.* (2001) Cancer and genomics. *Nature* **409**, 850–852.

Two articles about the application of genomics to the study and treatment of cancer.

Celis JE, Kruhoffer M, Gromova I *et al.* (2000) Gene expression profiling: monitoring transcription and translation products using DNA microarrays and proteomics. *FEBS Lett* **480**, 2–16.

Lakhani SR, Ashworth A (2001) Microarray and histopathological analysis of tumors: the future and the past? *Nature Rev Cancer* **1**, 151–157.

Reviews discussing the use of transcriptomics and proteomics to discover disease markers and in cancer diagnosis.

Funaro A, Horenstein AL, Santoro P *et al.* (2000) Monoclonal antibodies and therapy of human cancers. *Biotechnol Adv* **18**, 385–401.

A comprehensive account of the use of recombinant antibodies in cancer diagnosis and therapy.

Garrett MD, Workman P (1999) Discovering novel chemotherapeutic drugs for the third millennium. *Eur J Cancer* **35**, 2010–2030.

Gibbs JB (2000) Mechanism-based target identification and drug discovery in cancer research. *Science* **287**, 1969–1973.

Workman P (2001) Changing times: developing cancer drugs in genomeland. *Curr Opin Invest Drugs* **2**,1128–1135.

A series of reviews about cancer drug development, focusing on the role genomics has to play in drug discovery.

Hahn WC, Weinberg RA (2002) Modelling the molecular circuitry of cancer. *Nature Rev Cancer* **2**, 331–340.

Hanahan D, Weinberg R (2000) The hallmarks of cancer. *Cell* **100**, 57–70.

Jones PA, Baylin SB (2002) The fundamental role of epigenetic events in cancer. *Nature Rev Genet* **3**, 415–428.

Recent reviews on the molecular basis of cancer.

Various authors (2002) A Trends guide to cancer therapeutics. *Trends Mol Med* **8** (4 Suppl).

A collection of 12 excellent reviews on the development of small-molecule cancer drugs, plus a review of cancer gene therapy which is relevant to Chapter 8.

CHAPTER SIX

The large scale production of biopharmaceuticals

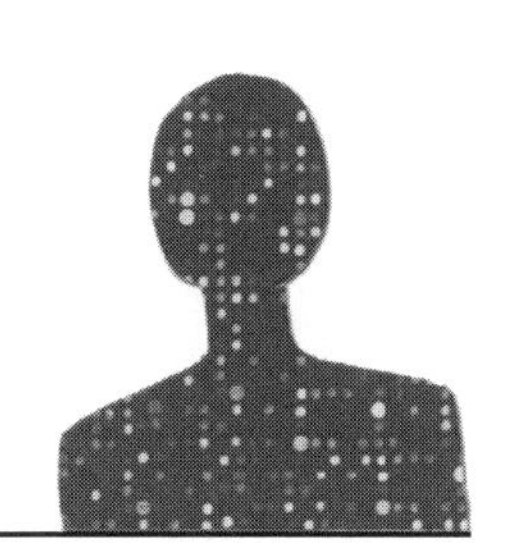

Overview

There are four types of biopharmaceutical: nucleic acids, proteins, viruses, and cells (Table 6.1). With the exception of some nucleic acids that are synthesized chemically, all biopharmaceuticals are produced by the large scale cultivation of microbial or animal cells (Table 6.2). The large scale culture of microorganisms is undertaken in fermenters that essentially are stirred tank reactors. Animal cells are much more fragile than microbial cells and cannot withstand the shear forces generated in a conventional fermenter. Nor are some of them capable of growth in suspension culture thereby necessitating the use of massed banks of roller bottles. Regardless of the method of culture, the objective is to maximize the synthesis of product and this is achieved by the use of appropriate vectors, expression systems, and culture conditions. The product then has to be extracted and purified and this stage is known as **downstream processing**. Because the final product is a

Table 6.1 Examples of the different types of biopharmaceutical.

Type of product	Examples
Nucleic acid	DNA vaccines Vectors for gene therapy Antisense oligonucleotides
Proteins	Therapeutic proteins (including antibodies) Diagnostic antibodies
Viruses	Bacteriophages as therapeutic agents Vaccines Vectors for gene therapy
Cells	Bacterial vaccines Cells for cell therapy

Table 6.2 The different hosts used for the production of biopharmaceuticals.

Production host	Product
Bacteria (principally *Escherichia coli*)	Therapeutic proteins Bacterial vaccines Bacteriophages for therapy Plasmid DNA for gene therapy and DNA vaccines
Yeasts (principally *Saccharomyces cerevisiae* and *Pichia pastoris*)	Therapeutic proteins Hepatitis B vaccine
Animal cells	Viruses as vaccines and vectors for gene therapy Therapeutic proteins Monoclonal antibodies for therapy and diagnosis Cells for cell therapy

pharmaceutical all the stages in its production have to be undertaken according to the principles of **good manufacturing practice** (GMP).

Examples of the many different therapeutic proteins now on the market are shown in Table 6.3. Of these biopharmaceuticals, monoclonal antibodies are unique in that they have so many uses and formats. As well as being used to treat cancer and various infections, they can be coupled with radioisotopes to generate *in vivo* diagnostics for imaging cancer. Furthermore, for the treatment of cancer the antibodies can be used alone or in combination with radioisotopes or selected toxins (see Chapter 5). Finally, antibodies can be of murine origin but, increasingly, humanized versions are being developed.

The generation of monoclonal antibodies

When a foreign macromolecule (**antigen**) is introduced into the circulatory system of a higher vertebrate it stimulates lymphocytes to produce antibodies that combine specifically with the macromolecule to facilitate its destruction or elimination. These antibodies are found in the globulin fraction of the proteins that circulate in the blood and hence are called **immunoglobulins**. All immunoglobulin molecules have a similar basic structure of two heavy and two light chains held together by disulfide bonds (Fig. 6.1). Regions of each chain are associated with particular functions. The **variable (V) regions** are responsible for antigen binding. Within the variable regions are relatively constant framework sequences that seem to act as a scaffold for the hypervariable regions also known as **complementarity-determining regions (CDRs)**, which are largely responsible for antibody specificity. Each different antibody has a different amino acid sequence and spatial arrangement and it is conceivable that every possible shape presented by an antigen can be accommodated by some antibody produced by the immune system. When an antibody is treated with papain, the heavy chain is cleaved once to produce three

Table 6.3 Examples of therapeutic proteins that currently are being marketed. Note that only a small number are being used to treat single gene disorders (see p. 91). Details of the nomenclature of therapeutic antibodies are given in Box 6.1.

Therapeutic protein	Clinical indication
Human insulin	Diabetes
Human growth hormone	Pituitary dwarfism
Hepatitis B vaccine	Prevention of hepatitis B infection
Interferon alpha	Hairy cell leukemia
Tissue plasminogen activator	Acute myocardial infarction
Erythropoietin	Anemia associated with renal failure
Interferon gamma	Chronic granulomatous disease
Granulocyte macrophage colony stimulating factor	Bone marrow transplant
Granulocyte colony stimulating factor	Chemotherapy-induced neutropenia
Interleukin 2	Renal cell carcinoma
Factor VIII	Hemophilia A
Human DNase	Cystic fibrosis
Glucocerebrosidase	Gaucher's disease
Interferon beta	Multiple sclerosis
Factor IX	Hemophilia B
Consensus interferon	Chronic HCV infection
Platelet growth factor	Chemotherapy-induced thrombocytopenia
Platelet-derived growth factor beta	Lower extremity diabetic ulcers
Tumor necrosis factor receptor linked to Fc portion of human IgG1	Rheumatoid arthritis
Glucagon	Hypoglycemia
Factor VIIa	Hemophilia
Monoclonal antibodies	
Gemtuzumab	Acute myeloid leukemia
Rituximab	Non-Hodgkin's lymphoma
Trastuzumab	Metastatic breast cancer
Palivizumab	Pediatric respiratory syncytial virus
Infliximab	Crohn's disease and rheumatoid arthritis
Basiliximab	Acute organ rejection in transplants
Daclizumab	Acute kidney transplant rejection
Edrecolomab	Colorectal cancer
Abciximab	Prevention of blood clots
Muromomab	Acute kidney transplant rejection

fragments: two antigen-binding fragments (Fab) and a constant region fragment (Fc).

When the immune system of an animal encounters a new antigen it responds by synthesizing a number of different antibodies depending on the number of antigenic determinants (**epitopes**) located on it. Following induction of antibody formation the animal can be bled and the serum fraction obtained. However, this serum will contain all the different antibodies produced in response to the antigen. Furthermore, these antigen-specific antibodies will be diluted with all the other antibodies present in the serum as a result of the animal's previous encounters with

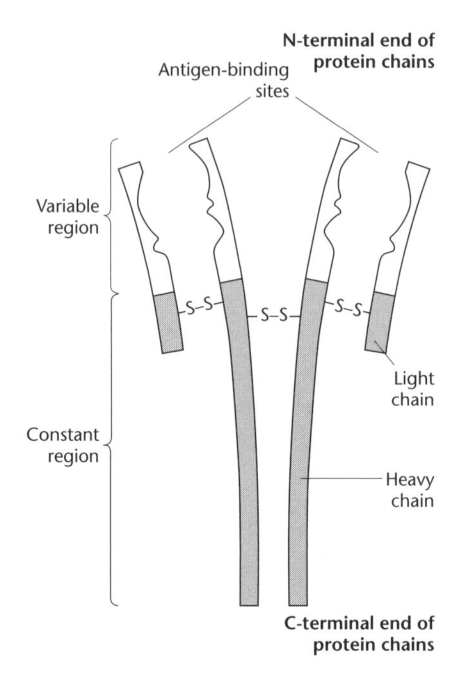

Fig. 6.1 Representation of an immunoglobulin molecule.

other antigens. Such polyclonal antisera cannot be used as biopharmaceuticals except in very specialized circumstances. Monospecific antibody species are required and these are known as **monoclonal antibodies**. They are synthesized by **hybridomas** that are produced by fusing antibody-producing mouse lymphocytes with an immortal mouse myeloma cell line (see p. 141). If all the hybridomas that occurred after a fusion were grown together then a polyclonal antibody mixture would result. Consequently, individual hybridoma cells are isolated and screened to find those that produce the monoclonal antibodies of interest.

Several types of human tumors express antigens on their surfaces that distinguish tumor cells from normal cells. This makes tumors potential targets for antibodies that recognize tumor-specific surface molecules. However, when mouse monoclonal antibodies are used for cancer therapy the patients often develop severe immune reactions. These occur because of the production of **human anti-mouse antibodies (HAMAs)**. Synthesis of HAMAs occurs, within 8–10 days and reaches a peak at about 25–30 days, precluding continued treatment. The ideal solution would be to use human monoclonal antibodies but the generation of human hybridomas has met with little success.

Genes encoding functional heavy and light chains of a desired mouse monoclonal antibody can be isolated from a hybridoma. Functional genes encoding heavy and light chain genes also can be isolated from human lymphocytes. By exchanging the variable regions (Fv) it is possible to create chimeric genes. When introduced into eukaryotic cells these genes express **chimeric antibodies** that

Box 6.1 The naming of monoclonal antibodies

The generic (nonproprietary) names of all drugs provide a clue to the molecule's chemical class and therapeutic class. The US Adopted Name (USAN) Council has adopted the following guidelines for naming monoclonal antibodies:

1 The suffix *-mab* is used for monoclonal antibodies and fragments.

2 The source of the antibody is designated within the name. The animal source is an important safety factor when considering the potential for the development of antibodies to the therapeutic monoclonal antibody in patients. The following are approved identifiers:

a = rat
e = hamster
i = primate
o = mouse
u = human
xi = chimera
zu = humanized.

3 The general disease state subclass is included in the nonproprietary name:

viral = *-vir-*
bacterial = *-bac-*
immunomodulator = *-lim-*
tumors – colon = *-col-*
tumors – melanoma = *-mel-*
tumors – mammary = *-mar-*
tumors – testis = *-got-*
tumors – ovary = *-gov-*
tumors – prostate = *-pr(o)-*
tumors – miscellaneous = *-tum-*
cardiovascular = *-cir-*.

4 A distinct, compatible syllable is selected as the beginning prefix in order to create a unique name.

5 If the product is conjugated to another chemical or is radiolabeled then recognition of this is made by the addition of a second word.

Examples of the use of this nomenclature are:
Imiciromab = a mouse monoclonal antibody used to detect myocardial infarction
Daclizumab = a humanized monoclonal antibody used to prevent organ rejection
Basiliximab = a chimeric monoclonal antibody used to treat organ rejection.

are about 75% human. Although less immunogenic than the original mouse antibodies, chimeric antibodies still can trigger significant antibody responses (HACA: human anti-chimeric antibody). Nevertheless, chimeric antibodies are being used to treat cancer, rheumatoid arthritis, and myocardial infarction.

The amino acids that actually make contact with the epitope on an antigen principally are located in the CDRs. It is possible to graft mouse CDRs onto a human antibody framework thereby creating a **humanized antibody** that is greater than 90% human. This is not an easy procedure but already a number of such antibodies are on the market. The ultimate goal is a fully human antibody and two techniques have been developed to achieve this. The first of these is phage display and this is described in Box 6.2. In the second, gene manipulation techniques are used to knockout the gene clusters in mice for the synthesis of heavy and light antibody chains and replace them with the human equivalents. Such mice produce human antibodies and by creating hybridomas from their lymphocytes it is possible to generate human monoclonal antibodies.

Radioimmunotherapy and diagnostic imaging

A growing number of anti-tumor antibodies are available for the treatment of different types of cancer. Patients initially respond well in that there is marked tumor shrinkage, but in a significant number of them the tumors become refractory after

Box 6.2 Human antibodies derived through phage display

As shown in Fig. B6.2, antibody variable regions are derived from several different gene segments. The human heavy chain variable region is formed by combining:

- 1 out of a possible 65 V_H segments
- 1 out of a possible 27 D segments
- 1 out of a possible 6 J segments.

This combination generates approximately 10,000 different heavy chain variable regions. The human light chain variable region is formed by combining:

- 1 out of a possible 40 V_L segments
- 1 out of a possible 5 J segments.

These individual segments can be isolated and cloned into an expression vector. In an antibody V_H and V_L are parts of different polypeptide chains. However, it has proven possible to express heavy and light chain variable regions as part of the same polypeptide chain and create functional antigen-binding proteins. These mini-antibodies are known as single-chain variable fragments (ScFv).

By fusing genes encoding ScFv to the coat protein gene of bacteriophage M13 it is possible to create a library of bacteriophages in which different phage particles express different ScFv molecules. This is the technique known as phage display. If the library of ScFv phages is exposed to an immobilized antigen, only those phages expressing a complementary ScFv will bind to the antigen. These phages can be isolated and enriched by propagation. The V_H and V_L genes of the selected ScFv are isolated from the phage genome and used to reconstruct a complete antibody, similar to the process of making a chimeric antibody.

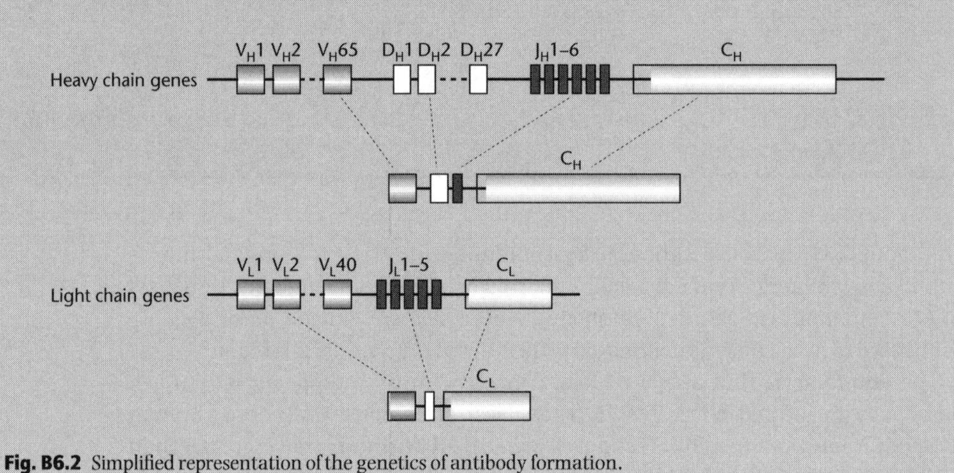

Fig. B6.2 Simplified representation of the genetics of antibody formation.

3–6 months' treatment. One way of minimizing this problem is to couple a radioisotope to the antibody. When the radiolabeled antibody binds to the tumor epitope the cells in the vicinity are subjected to localized radiation. This clearly is much less damaging to the patient than radiation that is applied externally. For the purposes of radioimmunotherapy the isotope of choice is ytrrium-90 since the energy of the particles it emits is sufficient to cause cell death but only in the immediate locality of the tumor. An alternative way of radiolabeling anti-tumor antibodies is to use either technetium-99m or indium-111. The emissions from these isotopes are sufficiently energetic that they can be detected outside the body with a special camera. This technique is known as gamma scintigraphy and can be used to precisely locate the tumor(s).

Other modified antibodies

Instead of using radioimmunotherapy, additional killing power can be given to an anti-tumor antibody by coupling to it a cellular toxin. Following binding of the immunotoxin conjugate to the tumor cell surface, it is transported into the cell where it does its damage. The toxins being evaluated for this purpose include ricin (irreversibly alters ribosomes), diphtheria toxin, *Pseudomonas* exotoxin, and calichaemicin. Cytokines such as tumor necrosis factor and interleukin 2 are also being coupled to antibodies to increase their tumor killing ability. In this case the idea is to increase the cytokine concentration in the vicinity of the tumor such that the immune system is triggered and starts destroying the tumor.

The large scale culture of microorganisms

Large scale cultures of microorganisms are often referred to as industrial fermentations. *Strictly speaking*, a fermentation is a biologic process occurring in the absence of oxygen but the term is now applied to any large scale cultivation of microbes even though most of them are aerobic. This is ironic for oxygen supply is the single most important factor limiting the efficiency of aerobic processes and modern fermenter design, and practice centers around the provision of adequate amounts of oxygen. Although facultative anerobes can grow in the absence of oxygen, their growth rates are lower and cell and product yields are greatly reduced.

A rapidly growing culture has a very high demand for oxygen. This oxygen must be dissolved in the growth medium so that it can interact with the membrane-bound electron transport system. The fundamental problem in supplying sufficient oxygen is that oxygen is very insoluble in aqueous systems, e.g. its solubility is only 7 mg/L at 35°C and 1 atmosphere. Furthermore, its solubility decreases as the temperature and the dissolved solute concentration increase. For small-scale culture, e.g. less than 1 L, it is possible to supply adequate amounts of oxygen by growing the culture in an Erlenmeyer flask that is agitated constantly. Agitation facilitates the transfer of oxygen from the gas phase in the flask to the liquid phase. Provided that the volume of the growth medium does not exceed 10–20% of the flask volume it is possible to achieve cell densities of 1–2 g dry weight/L before oxygen becomes limiting. Once the cell density increases beyond this level, oxygen utilization exceeds oxygen transfer to the liquid phase.

For large scale culture and/or high cell densities the oxygen demand of the culture can be met only by forced aeration. In practice, this is achieved by blowing air through the culture. The efficiency with which oxygen is transferred from air bubbles to the liquid phase principally depends on two functions: the surface area to volume ratio of the air bubbles and the residence time of the bubbles in the liquid. The smaller the bubbles, the greater the surface area to volume ratio and the greater the oxygen transfer. Similarly, the longer the bubbles remain in the liquid the greater will be the amount of oxygen that will diffuse from the bubble into the liquid. One way of decreasing bubble size is to introduce air through a sparger with multi-

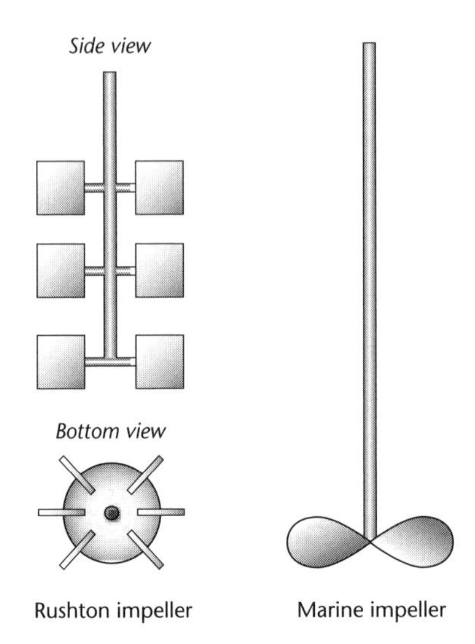

Fig. 6.2 Representation of a marine impeller and the six-bladed Rushton impeller.

Table 6.4 Some common features of fermenters.

Feature	Rationale
Baffles	Increase efficiency of oxygen transfer by increasing turbulence of the agitated culture medium
Antifoam control	Agitation and aeration of the culture medium can cause excessive foaming, particularly at high cell densities and/or if complex growth supplements (e.g. yeast extract) are used
pH control	The metabolism of most microorganisms results in a change in the pH of the culture medium. It is usual practice to monitor the pH of the fermentation broth continuously and to maintain a fixed pH by addition of acid or alkali
Temperature control	Initially the culture medium needs to be heated to the desired cultivation temperature. However, microbes produce heat as they metabolize substrates and so as the cell density increases the culture medium needs to be cooled
Addition ports	Provision must be made for the addition of the culture inoculum and media requirements

ple small orifices rather than through a single large-bore orifice. A second way is to agitate the culture broth vigorously with a stirrer. There are two additional advantages of agitation. The residence time of the bubbles is increased because they now follow a tortuous path rather than a straight path to the liquid surface. In addition, agitation facilitates oxygen transfer from the bulk liquid to the cell surface. Agitation is usually effected with a Rushton-type impeller (Fig. 6.2) since this is more efficient at bubble size reduction than a marine propeller. The number

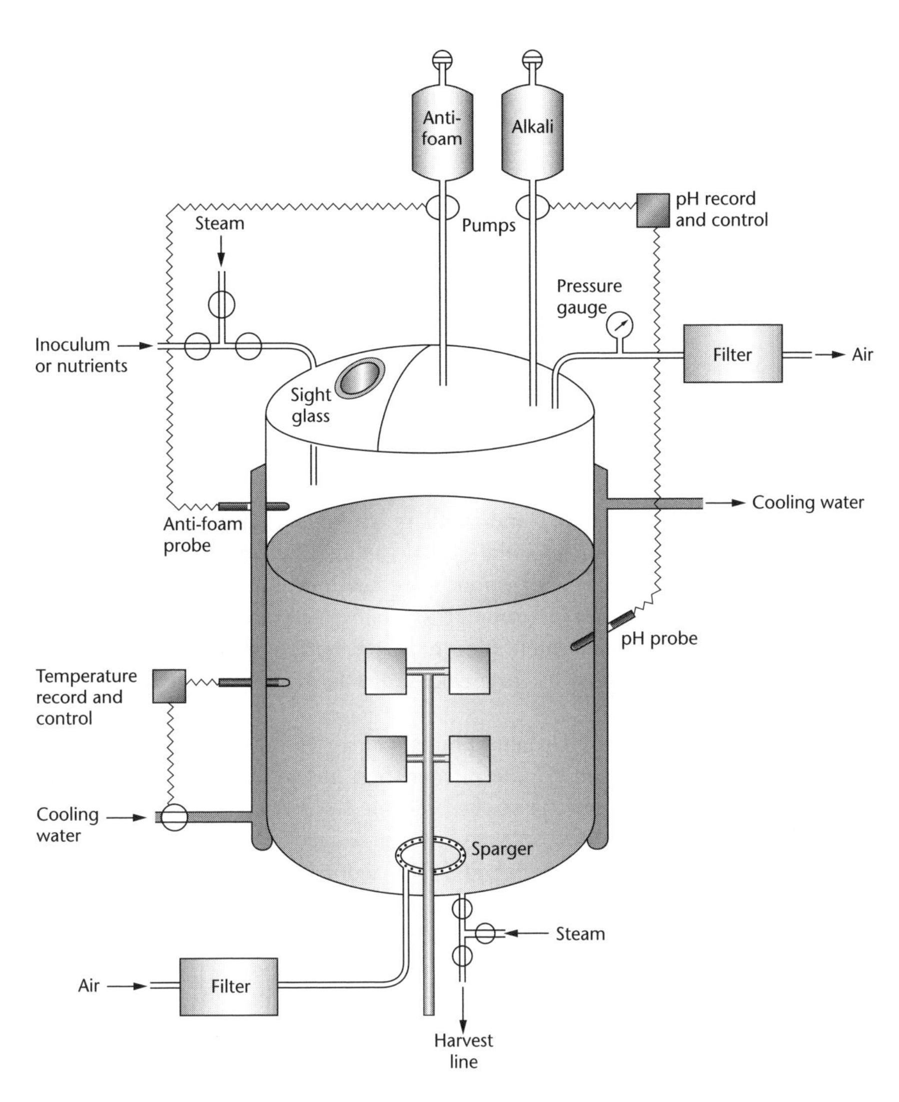

Fig. 6.3 Representation of a stirred tank reactor. For clarity no seal is shown between the agitator shaft and the fermenter body and baffles have been omitted.

of blades on the impeller and their dimensions relative to the fermenter dimensions greatly affect the efficiency of oxygen transfer.

From the above, it should be apparent that the basic fermenter consists of a closed vessel fitted with an air inlet and an agitator. However, many other features are required and some of them are outlined in Table 6.4. These requirements are embodied in the basic stirred tank reactor design as shown in Fig. 6.3. The material of fabrication is stainless steel and this must be of a high grade if it is not to corrode or leak toxic metal salts into the growth medium. A key requirement for a successful fermentation is aseptic operation: that is, no contamination of the culture should

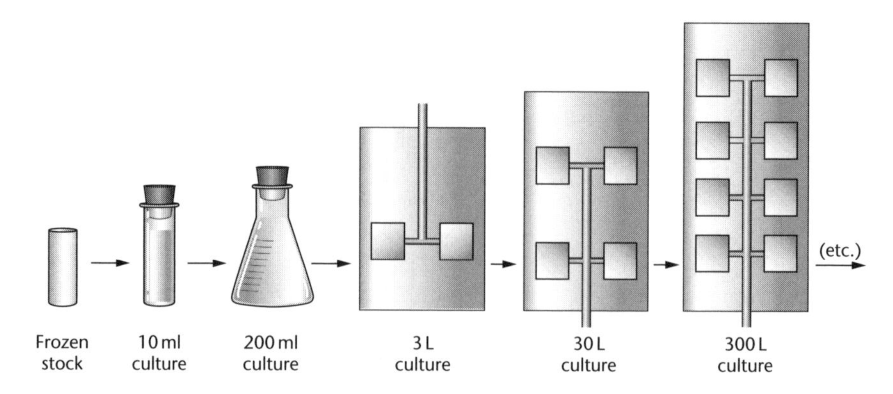

Fig. 6.4 A fermenter "train" used to provide a large inoculum for a production vessel.

occur. Thus the entire fermenter and ancillary equipment, as well as the growth medium, must be sterile before inoculation. In practice, the sterilization of all equipment is done *in situ* with the aid of steam. The growth medium is heat sterilized in the fermenter by passing steam through the cooling coils and jacket. In addition, the air supplied during the fermentation must be sterile and this is achieved by filtration. Finally, there must be no breaks in the fermenter that will allow the ingress of microorganisms.

The starting point for any fermentation is a clean vessel that is sterilized and charged with sterile medium. The fermenter then has to be inoculated and the size of the inoculum is generally of the order of 1–10% of the total volume of the medium. If it is any smaller there may be a prolonged lag period before growth commences and the fermentation period will be unduly prolonged. An Erlenmeyer flask culture can be used as the inoculum for a small fermenter with a capacity of 10–20 L. Once the fermenter volume increases beyond this it is necessary to prepare the inoculum in a smaller fermenter. For a very large production fermenter a fermenter train (Fig. 6.4) may be needed to provide the inoculum.

The large scale culture of animal cells

Before considering the methods for growing animal cells on a large scale it is necessary to understand the different types of animal cell that are used for the production of biopharmaceuticals. A piece of fresh tissue can be fragmented into its component cells by treatment with a proteolytic enzyme such as trypsin. If these cells are placed in a suitable growth medium they will attach themselves to the bottom of the vessel and begin to divide mitotically. A culture of this type, arising directly from a differentiated tissue, is referred to as a **primary culture**. Eventually the bottom of the culture vessel will be covered with a continuous layer of cells, often one cell thick, and referred to as a **monolayer**. The cells of primary cultures can be detached from the culture vessel and used to initiate **secondary cultures** by

Box 6.3 Immortalizing cells for genetic studies

The process of mapping a genetic defect and identifying and isolating the affected gene is a lengthy process (see Chapter 4) and can require substantial amounts of cellular material derived from patients. Often it is not practicable or desirable to keep going back to a patient for more tissue. The ability to immortalize lymphocytes solves this problem since only a small sample of blood can provide an indefinite supply of material. The way that this immortalization is done is summarized below.

A small sample of blood is withdrawn from the patient and the white blood cells separated from the red blood cells in a density gradient prepared from the polymer Ficoll. The white blood cells are washed and resuspended in culture medium at high density and Epstein–Barr virus is added. The infected cells then are grown in suspension culture and any B lymphocytes that have been transformed by the virus start to form visible clumps of cells. Eventually the T lymphocytes and untransformed B lymphocytes die off and only transformed cells (lymphoblastoid cells) survive. The lymphoblastoid cell line is expanded by further culture in larger volumes of fresh medium, divided into aliquots, and stored frozen. If, at a later date, more cells are required then these can be generated by subculture of an aliquot of the cell line in fresh medium.

reseeding them into fresh media. This process can be repeated a number of times but the cells eventually die after dividing about 40–80 times.

Not all animals yield primary cultures all of whose progeny die. In murine cell cultures a few cells become altered such that they acquire a different morphology, grow faster, and are able to start a culture from a smaller number of cells. Progeny derived from such exceptional cells have unlimited life and are designated **cell lines**. During repeated serial transfer cell lines can undergo extensive changes in their cultural properties. For example, the density at which division ceases may increase such that the cells grow in clumps rather than in monolayers and the cells may be irregularly oriented in respect to each other. Such cell lines are said to be **transformed** and are generally **neoplastic**, that is, they produce cancer if transplanted into related animals (see also Chapter 5). Lines of transformed cells also can be obtained from peripheral lymphocytes by infecting them with oncogenic viruses such as Epstein–Barr virus, a process known as **immortalization**. The ability to immortalize cells relatively easily is extremely valuable since it enables cells from individuals with specific genetic defects to be propagated indefinitely (Box 6.3). Whereas no human cell line has been shown unambiguously to originate from normal cells a number have been derived from human tumors. Apparently, the possession of the cancerous phenotype allows easier adaptation to growth in cell culture. The properties of a number of the cell lines used in the production of biopharmaceuticals are shown in Table 6.5.

The tendency of cell lines to change continuously on repeated cultivation necessitates that stocks of cells generally be maintained in the frozen state. The cells are mixed with additives such as glycerol or dimethyl sulfoxide to minimize cellular damage by ice crystals and are dispensed in ampoules and stored in liquid nitrogen. Cells maintained in this way remain viable for years and on thawing readily initiate new cultures.

One of the most crucial factors in achieving the successful cultivation of mammalian cells *in vitro* is the composition of the growth medium. To be satisfactory the

Table 6.5 Some of the cell lines used for the production of biopharmaceuticals.

Cell line	Species of origin	Tissue of origin	Growth in suspension
CHO	Chinese hamster	Ovary	Yes
BHK21	Syrian hamster	Kidney	Yes
NSO	Mouse	Myeloma	Yes
SP2/0	Mouse	Myeloma	Yes

Table 6.6 Criteria for a successful culture medium for animal cells.

1 The medium must provide all the nutritional requirements of the cell
2 The medium must maintain a pH value of 7.0–7.3 despite the production of acid, i.e. it must be adequately buffered
3 The medium must be isotonic with the cell cytoplasm
4 The medium must be sterile

medium must fulfill the criteria set out in Table 6.6. The growth media in common use are complex and generally not completely defined because of the presence of serum. The precise contribution made by the serum is not fully understood. Some components of the serum, e.g. α-globulins and hormones, may exert a beneficial effect by promoting the attachment and spreading of the cells and by stimulating cell division. The most popular source of serum is from fetal calves since this maintains better growth of cultured cells than serum from adult animals. Not surprisingly, fetal calf serum is prohibitively expensive and generally in short supply. Serum used in the production of biopharmaceuticals must be obtained from a reputable supplier who has a rigorous quality control program. Furthermore, the serum must be obtained from animals that have been certified as being free from bovine spongiform encephalopathy (BSE) prions. For these reasons a lot of effort has been extended on developing serum-free media for large scale use and such media now are used for the production of therapeutic proteins by suspension cultures.

The media used to culture animal cells are rich in nutrients so they are particularly susceptible to contamination with bacteria and fungi unless good aseptic technique is practiced. Because the loss of a large scale culture is very expensive antibiotics usually are added for additional protection. Another problem with cell cultures is contamination with mycoplasmas, a group of bacteria that lack cell walls. Most of the contaminating mycoplasmas are nonpathogenic and probably originate from the oropharynx of laboratory staff but some are derived from serum. They too can be inhibited with antibiotics.

The method used to grow animal cells on a large scale depends on whether they need to be attached to a solid surface (anchorage-dependent) or can grow in suspension. Growth in suspension is not very difficult since standard fermenters can be used. Fermenters are designed to run aseptically and have facilities for gassing and pH control. The stirrer motor may have to be geared down to ensure that a

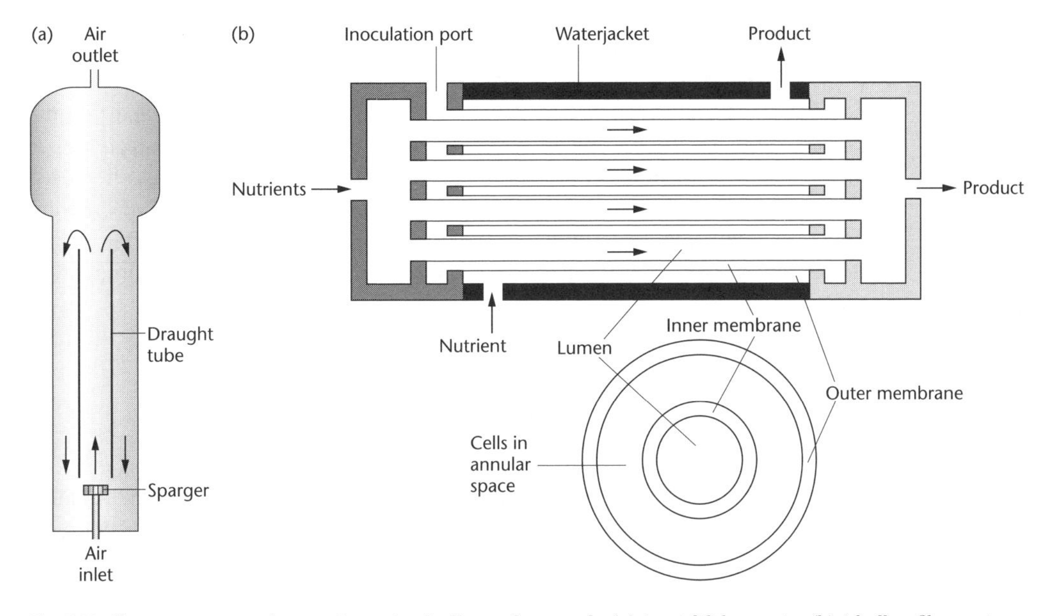

Fig. 6.5 Alternative system for growing animal cells on a large scale. (a) An airlift fermenter. (b) A hollow fiber system.

sufficiently low speed of stirring can be obtained to prevent undesirable shear effects on the cells. An alternative method of culturing the cells is to use an airlift fermenter (Fig. 6.5a) in which the sparged gas agitates the cells sufficiently to promote metabolite interchange without causing damage to the cells. More recently, a disposable culture system has been developed that makes use of the sterile bags used for media storage. These are inoculated and agitated on a rocking platform.

The growth of anchorage-dependent cells on a large scale is particularly difficult for a fermenter does not provide the requisite high surface area to volume ratio. A number of different methods have been developed and the choice of method depends partly on the use to which the cells will be put and partly on the capital that can be spent. The simplest system is the use of roller bottles that have a capacity of 500 ml to 50 L. These are filled with about one-twentieth of their volume of medium, inoculated with cells, and placed on their sides on slowly revolving rollers. As the bottles revolve the growing monolayer is alternately immersed in the liquid medium and exposed to the air in the bottle. Thus the cells are automatically washed with nutrients and exposed to oxygen and the movement of liquid medium ensures uniform distribution of its components. As can be readily envisaged, large scale roller bottle operation involves simple technology but is labor-intensive both in setting up the cultures and in the initial stages of downstream processing. A number of proprietary systems have been developed that eliminate the labor requirements of roller bottles. Essentially, these are stacks of flat chambers, each with a surface area of 632 cm^2, and with common fill and vent ports. A stack of 40 chambers is equivalent to 14 large roller bottles, but only one filling and emptying operation is required with each stack compared with 14 for the roller bottles.

Two alternative methods have been developed for growing anchorage-dependent cells on a large scale. The first of these makes use of microcarrier beads made from dextran or synthetic polymers. These beads are 50–200 μm in diameter and by adding them to a fermenter they greatly increase the surface area to volume ratio. Even better are porous microcarriers since they can support an even higher cell density and the matrix protects the cells from shear effects of the agitation used to maintain the beads in suspension. A completely different method of growing anchorage-dependent cells involves the use of hollow fiber technology (Fig. 6.5b). Although this method is used for small scale culture it is seldom used at production scale.

Expression systems

In large scale manufacturing the objective is to maximize the yield of product since higher yields improve the process economics and simplify the removal of contaminants. Therefore, molecular geneticists use their knowledge of vector replication and transcriptional control systems to develop strains in which the desired product is significantly overexpressed. For example, yields of plasmid DNA (for gene therapy) and therapeutic proteins of 250 mg/L and 2 g/L of *Escherichia coli* respectively have been reported. The corresponding figure for antibodies produced in cultured animal cells is 250–750 mg/L. However, the overexpression of a nonessential cellular component (the desired product) is a huge energetic drain on the producing cell and hence there is a strong selection for spontaneous mutants that produce much less plasmid or therapeutic protein. The longer the fermentation train (Fig. 6.4), the greater the number of cell divisions and the greater the risk that a low-yielding mutant will predominate in the final production vessel. For this reason, it is normal practice to use a strain in which gene expression can be controlled by environmental conditions. For example, when *Pichia pastoris* is the production host the desired protein is placed under the control of the methanol oxidase promoter. The organism is cultured in the absence of methanol until an appropriate cell density is reached in the final fermenter. Methanol is then added and there is a rapid burst of product synthesis.

Three principal factors affect the yield of recombinant products. The first of these is the copy number of the corresponding gene. If the gene is plasmid borne then gene amplification is achieved by using a plasmid with a high copy number. Many different cloning vectors with very high copy number are available but, as with overexpression of proteins, there can be a selection for mutants with lower copy number. In practice, it is possible to suppress or amplify the plasmid copy number by varying the environmental conditions within the fermenter, e.g. by varying the composition of the media and the growth temperature. It also should be noted that the larger the gene insert, and hence the larger the plasmid, the lower will be the copy number.

The second factor that can affect yield of product is the choice of promoter for gene expression. Many different high-expression promoters have been developed but most of these are suitable only for small scale laboratory experiments. To be useful for production purposes the promoter must be capable of being regulated by environmental conditions in a simple way that can be used on a large scale, e.g. the methanol oxidase example quoted above.

The third factor affecting product yield, and the one usually least understood by geneticists, is the method of culture. To maximize product yield it usually is necessary to optimize cell yield. Since more ATP is produced by oxidative metabolism than by fermentation it is essential that there is no oxygen limitation during culture. This explains the need for good aeration systems as described earlier (see p. 137). However, if nutrient concentrations are too high, the cells can switch to a fermentative mode even when excess oxygen is present. For example, at high glucose concentrations, *Saccharomyces cerevisiae* exhibits the Crabtree effect whereby it switches to alcoholic fermentation. Similarly, *Escherichia coli* only metabolizes the sugar as far as acetate. Consequently, sugars and other carbon sources usually are added in small quantities at regular intervals, a process known as batch feeding. Since fermentative metabolism also can suppress plasmid copy number and gene expression it is possible to use the growth conditions to minimize product formation, and hence strain instability, until the final production vessel.

Downstream processing

Once cell culture is complete it is necessary to recover the desired biopharmaceutical. At a minimum this will involve the separation of the cells from the fermentation broth (e.g. in the case of bacterial vaccines) but more usually includes purification of the product with or without cell disruption (Fig. 6.6). Some cells rapidly settle out of suspension once aeration and agitation have ceased but more usually cell removal is effected by centrifugation. Because of the volumes of fluid used in large scale manufacture it is necessary to use continuous flow centrifuges and some

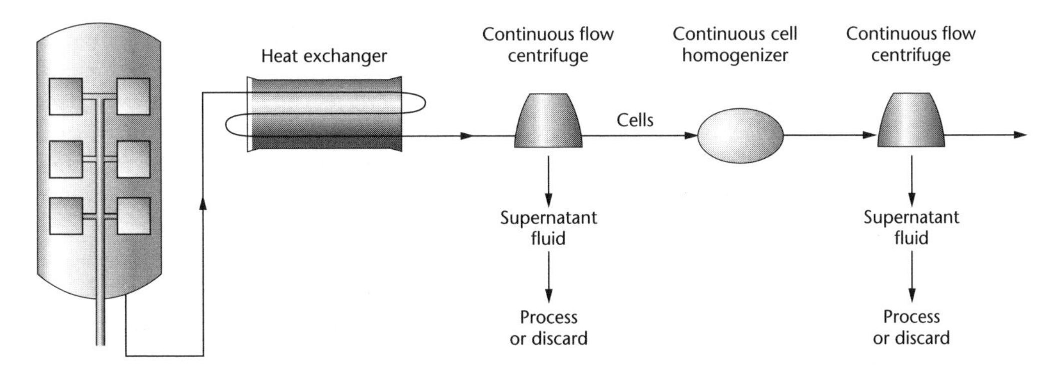

Fig. 6.6 Representation of typical operations in the processing of a fermentation broth. If the desired product is in the broth, the cells are discarded. If the product is inside the cells, the cells may need to be homogenized prior to product extraction.

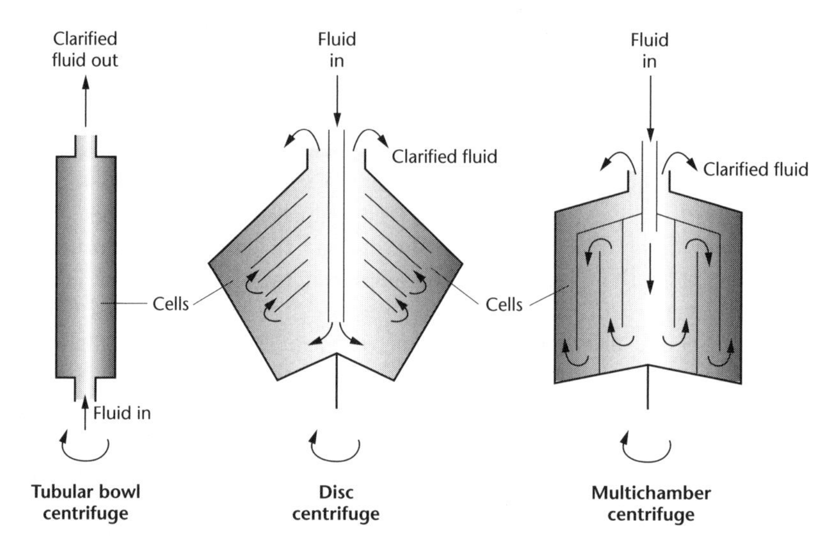

Fig. 6.7 Three commonly used designs of continuous-flow centrifuge.

examples are shown in Fig. 6.7. An alternative to centrifugation is ultrafiltration. The term ultrafiltration describes processes in which particles significantly greater in size than solvent molecules are retained when the solution is forced through a membrane of very fine pore size, usually less than 0.5 μm. Not surprisingly, with such fine pore sizes there is tremendous resistance to the passage of liquid and acceptable flow rates are achieved only by the application of pressure and by the use of high surface area to volume ratio. Proteins that have been secreted by the producing cells may or may not be retained by the membrane depending on the membrane pore size.

If the desired product is retained within the cell then cell disruption is necessary to effect its release. Although many methods of cell disruption are used in the laboratory most of them cannot be used on a large scale. Two that are suitable are high-pressure homogenization and the use of bead mills. In high-pressure homogenization cell suspensions move from a high-pressure environment to a low-pressure environment by passage through a fine orifice. The cells rupture by explosive decompression and the process is facilitated by their impaction at high velocity on components of the homogenizer (Fig. 6.8). High-speed ball mills are agitators consisting of a central shaft on which a set of circular discs is eccentrically mounted to form a helical array. The tank is filled with very small glass beads prior to charging with the cell slurry. It is the abrasion caused by the rapid motion of these beads that ruptures the cells.

After cell disruption the cell debris needs to be removed. As with removal of intact cells, the most widely used methods are centrifugation and ultrafiltration. Following this clarification step the desired product is subjected to a series of purification steps. The exact procedure used depends very much on the specific product and the likely contaminants (Table 6.7) but basically it occurs in three

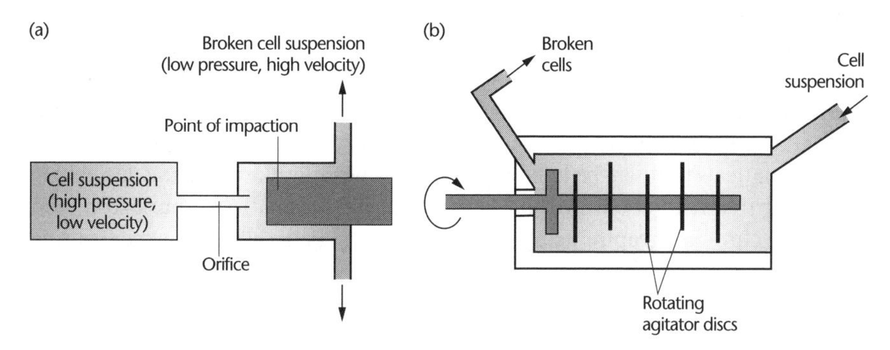

Fig. 6.8 Representations of equipment used for continuous disruption of cells (see text for details). (a) High-pressure homogenizer. (b) High-speed bead mill (beads not shown for clarity).

Table 6.7 Potential contaminants in biopharmaceuticals.

Source of contaminant	Type of contaminant
The host	Viruses (from animal cell hosts) Host-derived proteins and DNA Protein variants with altered glycosylation or other post-translational modifications Protein variants in which the N- or C-termini are not intact Endotoxins (from Gram-negative bacterial hosts)
The product	Denatured protein Conformational isomers Protein aggregates caused by misfolding or inappropriate cross-linking
The process	Growth medium components, especially serum-derived proteins Purification reagents including chromatography matrix materials Metals

stages. In the first stage the product is concentrated by capture on a matrix such as an ion exchange resin. In the next stage, which may involve numerous steps, the bulk of the contaminants are removed. In the final stage all trace contaminants and variant molecules are removed. It should be noted that some proteins when over-expressed in bacteria such as *E. coli* form inclusion bodies of aggregated protein. The protein in these inclusion bodies needs to be solubilized with denaturants such as 7 M urea or guanidinium hydrochloride and then renatured prior to purification. Some proteins such as epidermal growth factor renature very easily but with others it is impossible to regenerate a fully active protein.

Many of the purification procedures used by laboratory scientists are not suitable for large scale production. A good example is the purification of plasmid DNA. In the laboratory it is common practice to gently lyse cells with lysozyme and sodium dodecyl sulfate (SDS) to release the plasmid DNA and then to make a "cleared lysate" by pelleting the cells by centifugation. The supernatant fluid then is mixed with cesium chloride and ethidium bromide and subjected to density-gradient ultracentrifugation. Alternatively, the cell extract may be subjected to extraction with phenol and chloroform before ion exchange or reverse phase chromatography. These procedures are not suitable for the production of gene therapy vectors or DNA vaccines. Either they cannot be scaled up (density gradient centrifugation) or they use chemicals that are not acceptable for the production of pharmaceuticals that will be injected (SDS, phenol, chloroform). Instead, cells are broken by the methods shown in Fig. 6.8 and the plasmid DNA purified by ion exchange chromatography using reagents that are generally recognized as safe (GRAS).

Using gene manipulation to facilitate downstream processing of biopharmaceuticals

Many "foreign" proteins when synthesized in a new host cell such as *E. coli* are subject to protease attack at their N- or C-termini. These nibbled proteins often co-purify with the intact protein and their presence in the final product is not desirable. Synthesizing the desired protein as a fusion product can prevent nibbling. The fusion also can be used to facilitate purification. One of the earliest examples of purification fusions was the use of polyarginine tailing. In this method the gene sequence encoding the desired protein is extended by the inclusion at the 3′ end of a number of codons for arginine. When such genes are expressed the resultant proteins have a polyarginine tail or *tag* that makes them more basic. Upon ion exchange chromatography such proteins are separated from the bulk of the host cell proteins which are more acidic (Fig. 6.9). The polyarginine tail then is removed with the enzyme carboxypeptidase B, which can be immobilized for convenience. The de-tailed protein is rechromatographed on an identical ion exchange resin to separate it from any remaining contaminating proteins that will be more basic.

Examples of other purification tags that are in common use are glutathione-*S*-transferase, the MalE (maltose-binding) protein from *E. coli*, and multiple histidine residues. However, in each of these cases a proteolytic cleavage site has to be engineered between the desired protein and the tag. It is also possible to include in the tag a protein sequence that can be assayed easily. This can facilitate assay of the cloned gene product during purification when the regular assay is laborious or inconvenient.

Gene manipulation also can be used to minimize the sequestration of product into inclusion bodies as described in the previous section. One way of doing this is to engineer the host cell to overproduce a chaperon (e.g. DnaK, GroEL or GroES proteins) in addition to the protein of interest. Another method is to make minor changes to the amino acid sequence of the target protein. For example replacing one or more cysteine residues with serine can enhance solubility, although the resultant protein is not identical to the normal human form. The third method is derived from the observation that many proteins produced as insoluble aggregates in their native state are synthesized in soluble form as fusions with proteins such as thioredoxin, bacterioferritin, or the MalE protein.

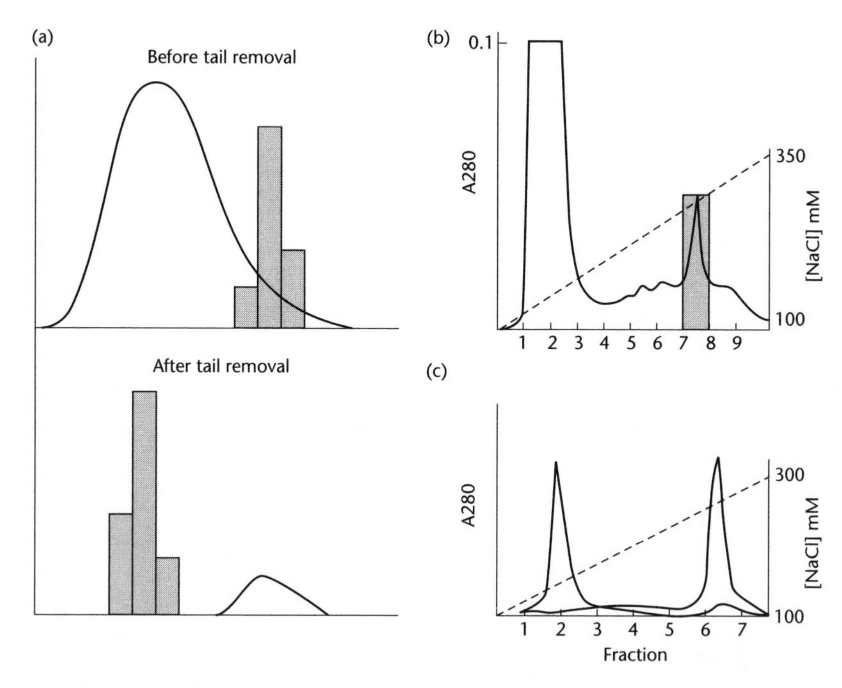

Fig. 6.9 The use of polyarginine-tailing to facilitate protein purification. (a) Representation of a hypothetical protein before and after enzymic removal of the C-terminal arginine residues. (b) Separation of polyarginine-tailed urogastrone from the bulk of the proteins in an *Escherichia coli* extract. (c) Chromatographic behavior of tailed and untailed urogastrone. The dashed line shows the salt gradient.

DNA for nonviral gene therapy is exclusively produced as plasmid DNA (pDNA) in *E. coli*. Characteristic of this production system is contamination of the pDNA with RNA and chromosomal DNA during purification and degradation by nucleases. To minimize these problems the bacterial cells are lysed with alkaline detergent, a process that precipitates chromosomal DNA and bacterial proteins. However, RNA still needs to be removed. One way of doing this is to engineer the gene for bovine pancreatic ribonuclease into the producer cell so that the enzyme is located in the periplasmic space. This enzyme withstands the alkaline lysis procedure and when the cell extract is neutralized, it quickly digests any RNA contaminating the pDNA.

The quality of biopharmaceuticals

The two key quality parameters of a biopharmaceutical are purity and potency. Given that all biopharmaceuticals are administered by injection, and often directly into the bloodstream, the potential contaminants of most concern are any adventitious microorganisms, bacterial endotoxins, DNA from mammalian cells that might

Box 6.4 The use of lectin microarrays to analyze the glycosylation of biopharmaceuticals

Glycoproteins are proteins in which carbohydrate is covalently linked to the polypeptide through the side chains of amino acid residues. The carbohydrate chains (**glycans**) of glycoproteins are classified as either *N*- or *O*-glycans, depending on the amino-acid residues to which they are linked. *N*-glycans contain the amino sugar *N*-acetylglucosamine linked to the side chain nitrogen of asparagines. *O*-glycans are linked to hydroxyl groups and when they contain *N*-acetylgalactosamine are linked to serine or threonine. Hundreds of different glycan structures have been isolated from glycoproteins but this diversity is generated from different combinations of just a few structural elements (Fig. B6.4a). More important from a biopharmaceutical point of view, is that anywhere from one to tens of glycan structures can be associated with a particular glycoprotein. Even different glycosylation sites within the same glycoprotein can carry different subsets of structures. The variation in glycan structures associated with the same glycosylation site in a population of glycoproteins with the same amino acid sequence is termed **microheterogeneity**. Glycoproteins with the same amino acid sequence but different glycans are known as **glycoforms**.

Different glycoforms of the same protein have different pharmacokinetic profiles. Thus it is important that during the development of the manufacturing process conditions are selected that favor the production of the most desirable glycoform(s). Furthermore, once the process enters production it is a regulatory requirement that there is batch-to-batch consistency in glycoform content. The difficulty in achieving these goals is that analysis of glycosylation patterns usually is very laborious and very lengthy. Some biopharmaceuticals, such as interferon beta, have a single glycan chain and only a few simple glycoforms have been observed. In this case analysis can be done by HPLC after release of the glycan chains from the protein. In other cases, such as erythropoietin, there are multiple glycosylation sites with complex glycans attached and many different glycoforms are known. The only way of properly analyzing these is to use a variety of mass spectrometry techniques and analysis of a single sample can take weeks. A recent innovation is the use of lectin microarrays for rapid glycoprotein analysis.

Lectins are sugar-binding proteins of nonimmune origin that agglutinate cells or precipitate glycoproteins. Generally, lectins bind to oligosaccharides rather than to single sugars and different lectins have different specificities. By determining which lectins are bound by a biopharmaceutical it is possible to elucidate the glycan structure. In practice, many different lectins are spotted onto a membrane to generate a microarray. The test protein (biopharmaceutical) is added and the different fluorescently labeled lectins are added to generate a lectin sandwich. The pattern of fluorescent labeling is determined (Fig.B6.4b) and algorithms are used to determine the glycosylation pattern. A corollary of this is that two identical samples of a biopharmaceutical

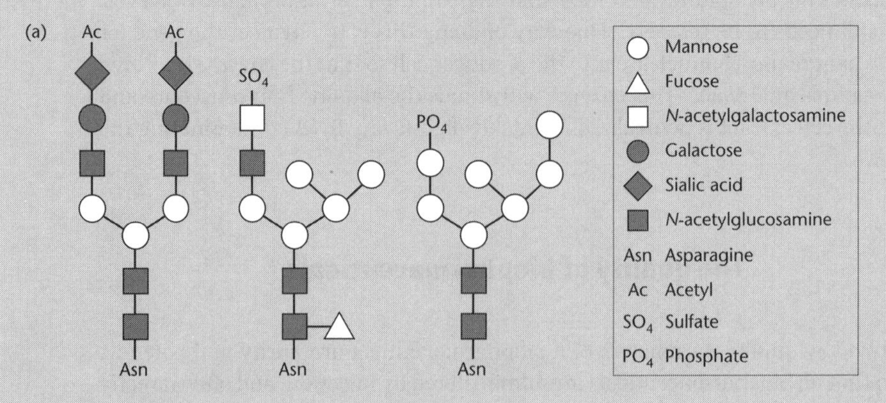

Fig. B6.4a Examples of some N-glycans.

will exhibit the same pattern in a lectin sandwich and this can be a useful quality control tool.

Figure B6.4b shows the glycosylation fingerprints produced with two different samples of a glycoprotein. Analysis of these fingerprints shows that the two samples are not identical. A key difference is the absence of terminal sialic acid in sample 1, since the presence of a terminal sialic acid residue can enhance the *in vivo* half-life of a protein that is used therapeutically. From the fingerprint it can be seen that both samples lack fucose, terminal GalNAc residues, and high mannose structures. Other information that can be obtained from the fingerprint, although not directly apparent, is the absence of hybrid glycans and truncated glycans, a low abundance of biantennary structures, and a high abundance of triantennary structures.

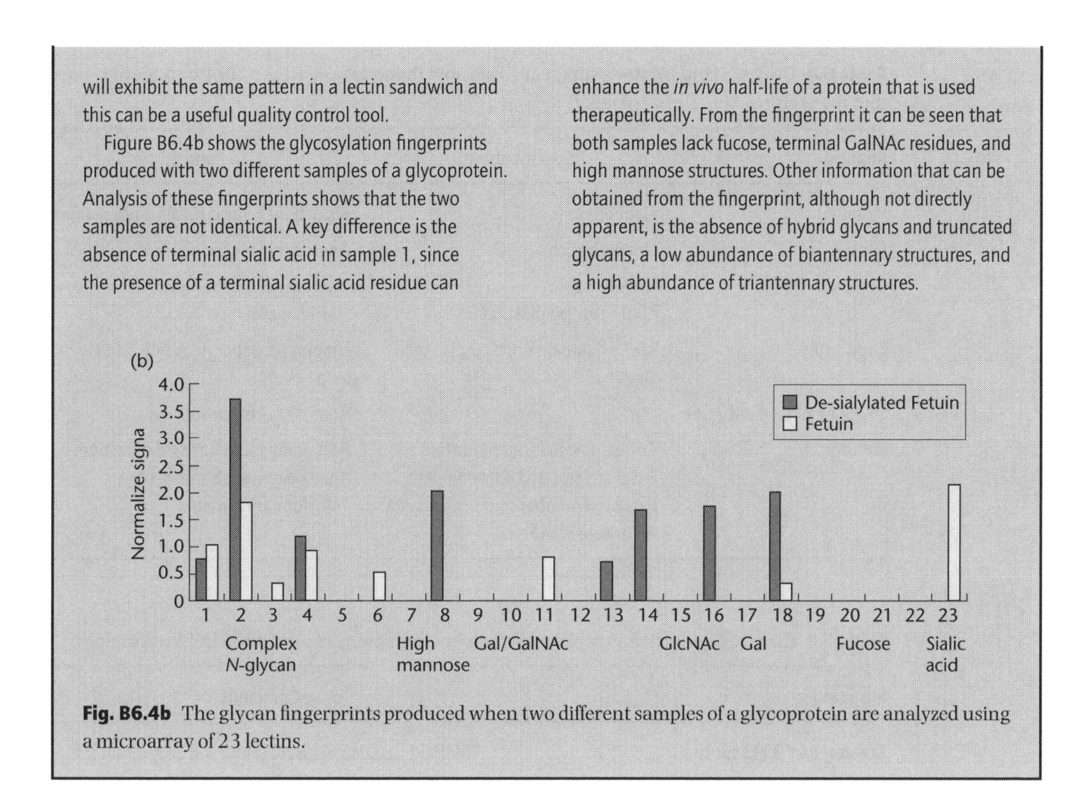

Fig. B6.4b The glycan fingerprints produced when two different samples of a glycoprotein are analyzed using a microarray of 23 lectins.

carry oncogenes, and any cell-derived viruses. However, contaminating proteins and variants of the target protein (e.g. degradation products or incorrectly folded protein) also are of concern. The exact quality standards required vary from product to product but the general requirements are shown in Tables 6.8 and 6.9. The finished product must meet these quality control standards but, more important, the purification procedure used must have been **validated**. That is, it must have been shown to remove these contaminants completely when they were deliberately added to the starting material.

The potency of biopharmaceuticals can be influenced by their primary structure (amino acid sequence), their shape and folding (secondary and tertiary structures), and the relationship between the protein molecules themselves (quaternary structure). Endogenous proteins can be heterogeneous at each level. For example, natural gamma interferon exists as six variants with differing C-termini. In addition, many proteins undergo post-translational modification such as glycosylation (Box 6.4). For proteins such as erythropoietin that have multiple glycosylation sites tens of different glycoforms may exist naturally. The significance of glycosylation for potency is best illustrated by interferon beta. The natural molecule is glycosylated at a single site and has a specific activity of 3×10^8 units per milligram. When it is produced in *E. coli* it is unglycosylated and the specific activity is only 3×10^6 units per milligram. However, by changing the cysteine residue at position 17 to a serine residue the full specific activity can be obtained with material

Table 6.8 Quality standards for a supercoiled plasmid that is to be used as a DNA vaccine or a gene therapy vector.

Impurity	Specification	Assay procedure
Proteins	Undetectable	Bicinchoninic acid (BCA) protein assay
RNA	Undetectable	Agarose gel electrophoresis
Genomic DNA	Undetectable <0.01 μg/μg plasmid DNA	Agarose gel electrophoresis Southern blot
Endotoxins	<0.1 endotoxin units/μg plasmid DNA	Limulus amoebocyte lysate (LAL) assay
Linear or relaxed plasmid	<5%	Agarose gel electrophoresis
Identity	Correct fragmentation pattern Expected size and supercoiling Expected number of transformants relative to standard	Restriction endonuclease digestion Agarose gel electrophoresis Transformation assay

Table 6.9 Quality standards for therapeutic proteins made using recombinant DNA technology.

Standard	Comment
Greater than 95% purity	Methods used to measure purity will depend on the particular protein but can include polyacrylamide gel electrophoresis and HPLC
Microheterogeneity below specified level and/or consistent pattern of microheterogeneity	Methods used will depend on expected heterogeneity but can include HPLC and mass spectrometry
Contaminating DNA less than 10 pg per dose	For products made in animal cells with/without vectors derived from oncogenic viruses the acceptable levels might be much lower
Endotoxin below specified level	This is principally for products made in *E. coli*
Chemicals used in purification below a specified level	Methods used depend on likely contaminants
Absence of microorganisms	Sterility tests on a variety of media
Specific activity above a minimum level and preferably the maximum achievable	Specific activity must be measured in one or more relevant biologic assays

produced in *E. coli*. It should be noted that determining the potency of vaccines can be problematic when suitable animal models are not available since the key property of a vaccine is its ability to generate an appropriate immune response in humans.

Differences in glycosylation may also have an influence on the pharmacokinetics and the immunogenicity of a biopharmaceutical. For example, removal of terminal sialic acid residues from the sugar chains of erythropoietin reduces the *in vivo* half-life from 8 hours to 5 minutes! The removal of sialic acid exposes galactose residues and these bind to galactose-recognizing receptors in the liver resulting in a rapid loss of asialo-erythropoietin from the systemic circulation. It is worth noting that when cells are engineered to overproduce a therapeutic protein the level of expression of that protein might be such that the glycosylation apparatus cannot keep pace. In such cases this could mean that novel glycoforms are produced that the body recognizes as foreign. This in turn could lead to unwanted immune responses and such adverse effects have been noted with some brands of erythropoietin.

Good manufacturing practice

The quality of a pharmaceutical product cannot be ensured solely by examining in detail a small number of units taken from a completed batch. The problem is that a low level or uneven distribution of a contaminant may not be detected by quality control tests on the finished product. Significant examples of this problem occurred in the 1960s when a number of hospital deaths were traced to microbial contamination of solutions administered intravenously. These solutions had been manufactured commercially, but not sterilized properly, and the quality control (QC) tests failed to detect any residual contamination. These incidents led to the development of the code of good manufacturing practice (GMP) that now is standard practice in the pharmaceutical industry (Box 6.5). Essentially, GMP demands that the product is manufactured from raw materials of the highest standard in a facility of the proper design by staff with the appropriate training using a procedure that is capable of reproducibly yielding a final product of the correct quality that will pass all QC tests.

A key part of GMP is process validation, i.e. proving beyond doubt that the process can regularly and reliably deliver material that is suitable for its intended use. Since biopharmaceuticals are administered by injection they must be free of microorganisms, prions, nucleic acids (except vectors), oncogenes, and any materials used in the purification of the product. Clearly, the starting point is to use materials of the highest quality and of known provenance. For example, by using fetal calf serum that comes from herds free of BSE there is no danger that the product will be contaminated with BSE and hence no possibility that the recipient will contract Creutzfeld–Jakob disease. This illustrates a key point: it is much easier to use material free of a potential contaminant than to prove that a process can reliably remove the contaminant were it to be present. Similarly, if the level of extraneous contaminants is kept to a minimum then there is a much higher probability that the manufacturing process will remove these contaminants. Thus, by working in premises with very high standards of hygiene, contaminating microbes will be kept to a minimum and for biopharmaceuticals the greatest risks will be from the production host.

For biopharmaceuticals produced in bacteria and yeast the principal contaminants of concern are endotoxins (Gram-negative bacteria only), contaminating proteins and, if the product is a therapeutic protein, structural variants of the product. Developing processes that eliminate these contaminants is not too difficult. In this regard, it is worth noting that as expression levels of product increase the less one

Box 6.5 Definitions used in the manufacture of pharmaceuticals

Manufacture

This is the complete cycle of production of a medical product. This cycle includes the acquisition of all raw materials, the processing of raw materials into a final product, and its subsequent packaging and distribution.

Quality assurance

This term refers to the sum total of the arrangements made to ensure that the final product is of the quality required for its intended purpose. It consists of good manufacturing practice (see below) plus factors such as original product design and development.

Good manufacturing practice (GMP)

This sometimes is referred to as good pharmaceutical manufacturing practice (GPMP). It comprises that part of quality assurance that is aimed at ensuring that a product is **consistently** manufactured to a quality appropriate to its intended use. GMP requires that the manufacturing process is fully defined before manufacturing commences and that the necessary facilities are provided. In practice, this means that:

- Personnel are adequately trained
- Suitable premises and equipment are employed
- Correct materials are used
- Approved procedures are adopted and no changes can be made without written authorization
- Suitable storage and transport facilities are available
- Appropriate records are kept and there is a complete audit trail.

Quality control

This refers to that part of GMP that ensures that at each stage of manufacture the necessary tests are made and that the product is not released until it has passed these tests.

In-process control

This comprises any test on a product, the environment or the equipment that is made during the manufacturing process.

has to worry about contaminants. For example, removing contaminating proteins is much easier if the desired protein is synthesized at 30% of total protein than when it is 3%. Validating processes involving animal cell culture is much harder for there is a need to show that the product is free of oncogenic viruses and oncogenes. Since the production hosts are transformed cell lines this means demonstrating conclusively that the purification process would remove these contaminants should they be present in the starting material.

Alternative production systems

All the discussion on this chapter has been on the production of biopharmaceuticals by large scale culture of microbial or animal cells. This is the way that biopharmaceuticals are made today – and it is an expensive one! As long as the prices that biopharmaceutical companies can achieve for their products are maintained there

Table 6.10 Relative costs of different production systems excluding capital costs.

System	Cost ($/g)
Chinese Hamster Ovary cells	300
Chicken eggs	2
Goats' milk	2
E. coli	1
Transgenic plants	0.1

is no great pressure to change the system. Nevertheless, a number of alternative systems have been developed experimentally. These include the production of therapeutic proteins in the milk of farm animals, in chicken eggs, and in plants, a methodology referred to as "pharming." Development of these products is sufficiently far advanced that provisional cost estimates can be made. The big advantage of "pharming" is reduced costs (Table 6.10). However, there also are disadvantages. Generating transgenic farm animals that produce the desired protein is not an easy task and there still will be the worry about oncogenic viruses and prions. Producing transgenic plants is, by comparison, much easier and there is no risk of transmission of human pathogens. However, fields of transgenic plants do attract environmentalists who have no qualms about destroying crops, and this does not fit with the tightly controlled manufacturing that regulatory agencies demand. Time will tell if these production systems get adopted on a large scale.

Further reading

POGM: A detailed description of plasmid copy number control, the generation of high expression vectors and the use of specialist vectors to facilitate purification is presented in Chapters 4 and 5. The production of proteins by "pharming" is described in Chapter 14 and builds on material presented in other chapters.

Baker KN, Rendall MH, Patel A *et al.* (2002) Rapid monitoring of recombinant protein products: a comparison of current technologies. *Trends in Biotechnology* **20**, 149–156.
This review covers a topic not included in this chapter.

Borrebaeck CAK, Carlsson R (2001) Human therapeutic antibodies. *Curr Opin Pharmacol* **1**, 404–408.
This review complements and expands on the material covered in this chapter.

Chu L, Robinson DK (2001) Industrial choices for protein production by large scale cell culture. *Curr Opin Biotechnol* **12**, 180–187.
This review complements and expands on the material presented in this chapter.

Crommelin DJA, Sindelar RD (eds) (2002) *Pharmaceutical Biotechnology*, 2nd edn.
This is an excellent textbook that expands on the topics described in this chapter and includes many other topics such as formulation, pharmacodynamics and pharmacokinetics, sterilization etc. not covered in this chapter or elsewhere.

Ferreira GNM, Monteiro GA, Prazeres DMF, Cabral JMS (2000) Downstream processing of plasmid DNA for gene therapy and DNA vaccine applications. *Trends Biotechnol* **18**, 380–388.

Levy MS, O'Kennedy RD, Ayazi-Shamlou P, Dunnill P (2000) Biochemical engineering approaches to the challenges of producing pure plasmid DNA. *Trends Biotechnol* **18**, 296–305.

These are two excellent reviews that present the issues of large scale plasmid purification in different but complementary ways.

Fischer R, Twyman RM, Schillberg S (2003) Production of antibodies in plants and their use for global health. *Vaccine* **21**, 820–825.

This review includes a number of successful case studies where antibodies have been produced in plants.

Middleberg APJ (2002) Preparative protein refolding. *Trends Biotechnol* **20**, 437–443.

This is an excellent review of a topic that seldom is discussed in the literature.

C H A P T E R   S E V E N

Genomics and the development of new chemical entities

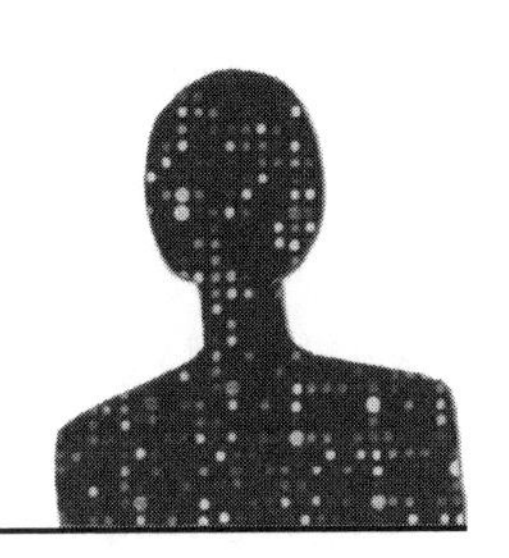

Introduction: how drugs are developed

Before the advent of the modern pharmaceutical industry, effective drug treatments were discovered in an arbitrary way. Usually this occurred by observing the effects of ingestion of plant extracts and good examples are aspirin (an anti-inflammatory agent) and quinine (an anti-malarial agent). Today, the risks of giving humans substances with unknown pharmacologic effects are not acceptable and the development of new drugs follows a standard and well-tested path (Fig. 7.1). Two ingredients are essential to the drug discovery process: a test system, or a battery of tests, and chemicals. The tests are usually biochemical assays for a protein (or **target**) whose function is implicated in a particular disease. The objective of the drug delivery process is to find either a chemical that suppresses the activity of the target (i.e. an **antagonist**) or one that increases its activity (an **agonist**).

Once one or more chemicals are identified that interact with the target in the desired way (**hits**) they are subjected to successive rounds of chemical modification (**optimization**) with the objective of identifying improved molecules (**leads**) with optimal biochemical properties *in vitro* and potency *in vivo*. This research phase is followed by a preclinical development phase. At this stage the lead molecule is formulated in such a way that its stability will be maintained and it will reach the target tissue in the body of test animals and humans. The adsorption, distribution, metabolism, and excretion (ADME, pharmacokinetics) of the drug will be determined following administration to test animals and its toxicity also will be determined. Toxicity tests take a number of forms. Acute toxicity tests are carried out by giving progressively increasing doses of compound to animals until an end-point (death) or an arbitrarily large limit has been reached. Chronic toxicity tests involve repeated dosing of the test compound and are designed to determine the effects of taking the drug for long periods. If it is intended that the drug be taken for prolonged periods then carcinogenicity testing will form part of the chronic toxicity testing. Reproductive toxicity testing is undertaken to ensure that the fetus is not adversely affected if pregnant females take the drug.

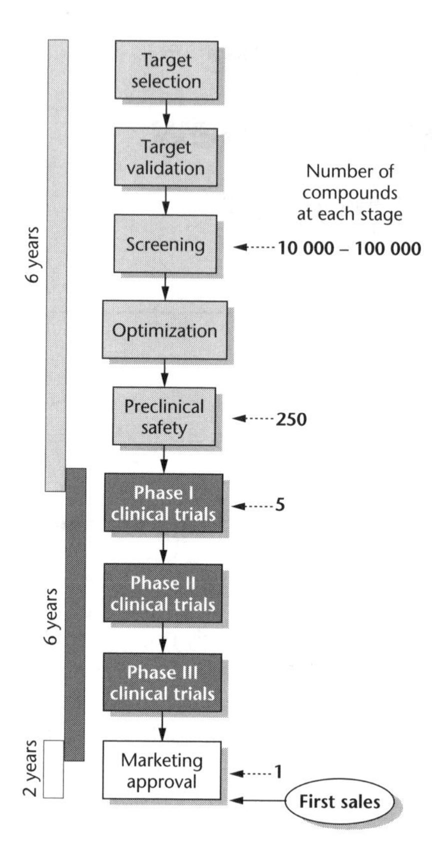

Fig. 7.1 The drug discovery process.

If the lead molecule performs satisfactorily in all the above tests it will proceed to clinical trials. These clinical trials proceed in four well-characterized phases. In phase I the drug candidate is administered to healthy human volunteers. There are two objectives of phase I studies: to assess the safety of increasing doses and to obtain basic pharmacokinetic information. The objective of phase II studies is to determine if the drug shows promise in one or more clinical indications and a small group of patients is exposed to the drug for the first time. The trials are designed to define the most suitable dosage schedule and to give an estimate of clinical efficacy in relation to the concentration of the drug and its metabolites. Phase III trials involve very much larger numbers of patients but only commence if there is evidence that:

- An adequate degree of efficacy exists.
- The risk profile of adverse reactions appears to be acceptable in terms of demonstrated efficacy.

There are two objectives of phase III trials. First, to generate data on efficacy and safety that will satisfy regulatory authorities such that permission is given for the drug to be marketed. Second, to generate information that will enable practicing doctors to utilize the drug effectively. Phase IV trials occur after the drug is launched on the market and are designed to determine drug efficacy in prolonged use on a much larger population of patients.

Traditionally, the conversion of a hit into a lead compound occurred via close partnership between chemists and biologists. Molecules, which were known to interact with the target, were modified in various ways and re-tested. Based on the results obtained attempts were made to determine structure–activity relationships (SARs) and further cycles of modifications to the molecule were undertaken until the biologists felt that they had a molecule with optimal properties that could be taken forward as a lead compound. The average annual output of an organic chemist is 50–100 compounds, depending on the complexity of the chemistry involved, and management would decide how many chemists would be employed per project per year. A reasonable time for lead selection if done as described above has been 1–2 years. The limiting factor generally has been the initial synthesis stage and the exploration of the numerous parameters associated with SAR development.

Starting about 15 years ago a new paradigm was developed for the early stages of drug discovery. This change was brought about by two developments. The first of these was the development of high-throughput assays that were amenable to simple automation and could be run in their thousands or tens of thousands every day. The second change factor was the development of methods for simultaneously synthesizing large numbers of chemical compounds that could be used to feed the high-throughput screens. The problem then became a lack of suitable new targets. With the advent of genomics, the identification of potential new targets no longer is rate limiting (see Chapters 3 and 4). Rather, the problem is one of validating the likely clinical utility of these targets.

The complete process of developing a new drug can be viewed conceptually as occurring in a filter funnel (Fig. 7.2). The identification of a lead molecule takes place in the cone and the preclinical safety testing and the clinical trials occur in the stem. With the advent of high-throughput screening, new chemical synthesis methodologies and new methods of target identification, the cone has become larger but the stem has become an even greater bottleneck. Given that 90% of the costs of new drug development occurs in the later stages it would be economically beneficial if some kind of prescreening could be undertaken to eliminate those drugs that might fail during safety testing and clinical trials. Fortunately, genomics is beginning to make contributions in these areas.

High-throughput screening

For a chemical to be active as a pharmaceutical it needs to bind selectively to one or a few cellular proteins, i.e. the target. Conceptually, therefore, an *in vitro* assay can be reduced simply to an ability to detect binding of test ligands to the target protein. If the target protein has enzymic activity then a better test is the inhibition of that activity. If large numbers of assays, e.g. thousands per day, are to be undertaken then the ideal binding assay is one in which there is no need to separate bound ligand from free ligand. Such an assay is said to be homogeneous. A number of

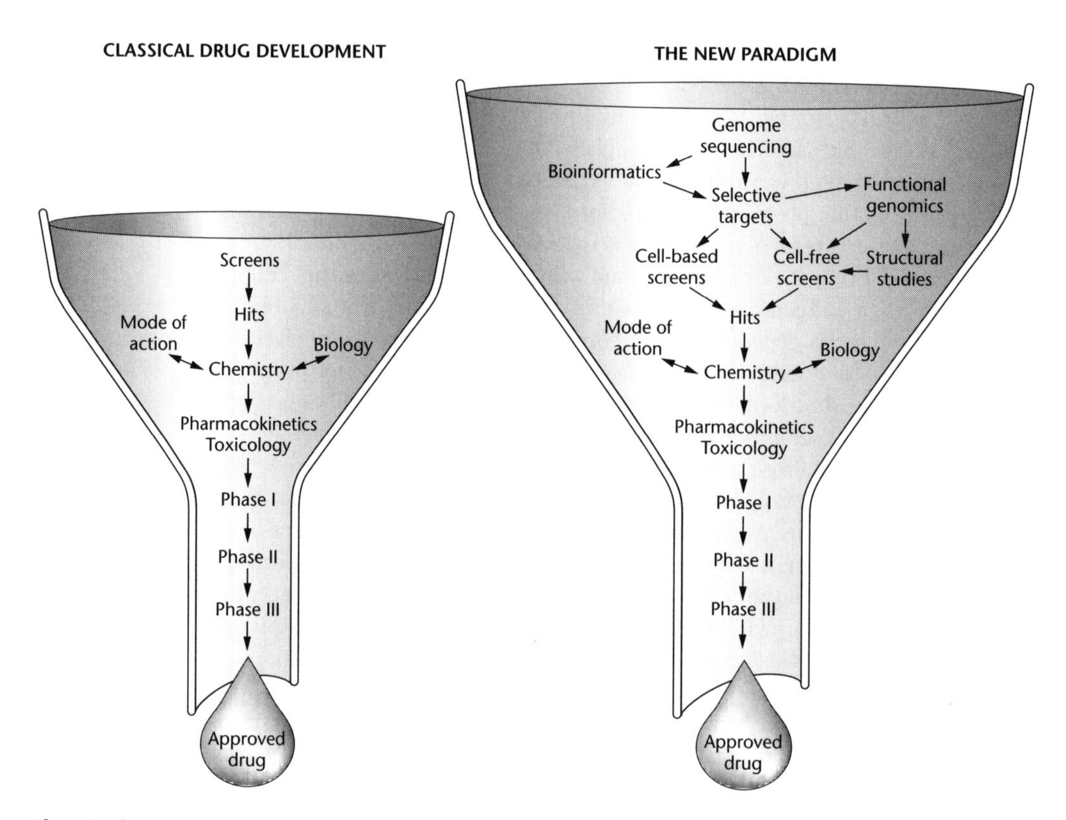

Fig. 7.2 Classical drug development versus the new paradigm.

homogeneous assay formats have been developed but only two generic ones will be described here: optical biosensors and scintillation proximity assays (SPA).

In the former, the target molecule is bound to the surface of a biosensor chip and then the test ligand is added. Any binding that occurs can be monitored in real time using an optical detection principle known as surface plasmon resonance (SPR). The SPR response reflects a change in mass concentration at the detector surface as molecules bind or dissociate from the sensor chip. The material of the biochips is amenable to the immobilization of a wide range of targets including DNA, enzymes, antibodies, and receptors, as well as more complex molecules such as liposomes and platelets.

With the SPA technology, the target molecule is bound to fluomicrospheres that act as scintillants; that is, the beads will give off light if a radioactive substance comes into close proximity. In practice, this only happens if the radiolabeled material binds to the target molecule (Fig. 7.3). For high-throughput binding assays one needs to start with a radiolabeled ligand that is known to interact with the target molecule. Test ligands then are added and any that diminish light output from the scintillant beads must be competing for binding to the target molecule. The beauty of the SPA technology is that it can be adapted readily to a number of different formats including enzyme assays (Box 7.1).

There are a number of problems associated with simple ligand-binding assays like those described above. For example, accessory factors that are essential for a

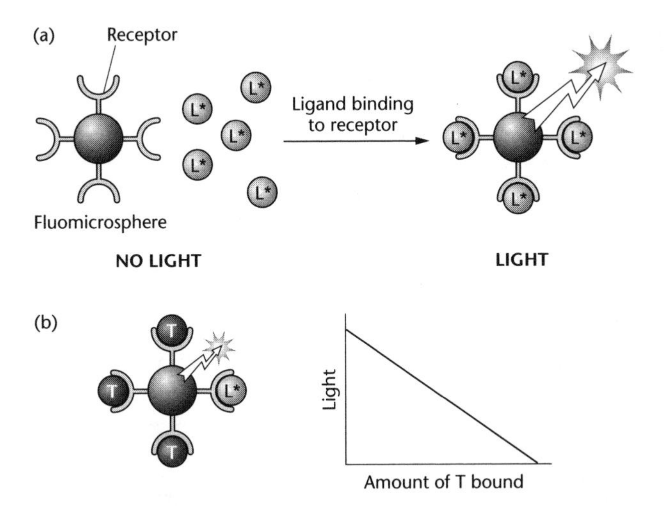

Fig. 7.3 The principle of scintillation proximity assays. (a) The binding of a radioactive ligand (L*) to the immobilized receptor results in the emmission of light. (b) An unlabeled test substance (T) is also capable of binding to the immobilized receptor and hence competes with L*. The graph shows that the higher the concentration of T, the less of L* that binds to the receptor and the lower the light emitted.

pharmacologic response might be absent, or present in altered ratios, in ligand-binding assays with purified proteins. Also, simple ligand-binding assays are biased for the detection of compounds that bind to a target at the same binding sites as the labeled ligand, and often to the same conformational state of the receptor. These limitations can be overcome by the use of functional cell-based assays (Fig. 7.4). They do not select for specific sites or modes of ligand interaction. Rather, they are capable of identifying novel compounds with different sites and modes of interaction.

Although cell-based functional assays have advantages over ligand-binding assays it is important that they are practical to perform and can be run in high-throughput mode. In particular, the assay platform must be adaptable so that it works with a large variety of genomic targets. Suitable formats have been described for targets as diverse as nuclear receptors, G-protein-coupled receptors, ion channels, and protein kinases (Table 7.1).

A different use of cell-based screens is in the discovery of new antibiotics. All the clinically useful antibiotics were identified originally as agents that possessed antibacterial activity. Such broad cell-based screening does not identify the biochemical target of a potential lead, thereby hindering attempts to optimize the compound on the basis of structure–activity relationships. Consequently, the pharmaceutical industry changed its antimicrobial discovery activities from inhibition of cellular growth to inhibition of specific biochemical targets. Although many potent enzyme inhibitors have been identified in this new approach none has been suitable as an antibiotic. Consequently, there now is a move back to cell-based assays but this time they are target-specific.

The starting point for the new antibacterial assays is the generation of strains that are more sensitive for specific enzyme inhibitors than for antibacterial activity

Box 7.1 Scintillation proximity assays

(1) The scintillation proximity assay (SPA) format is extremely versatile and can be used in high throughput mode to monitor ligand binding and a wide variety of enzyme reactions. Two types of ligand binding assay are used widely, binding to antibodies and binding to receptors, and both are easily converted into scintillation proximity assays.

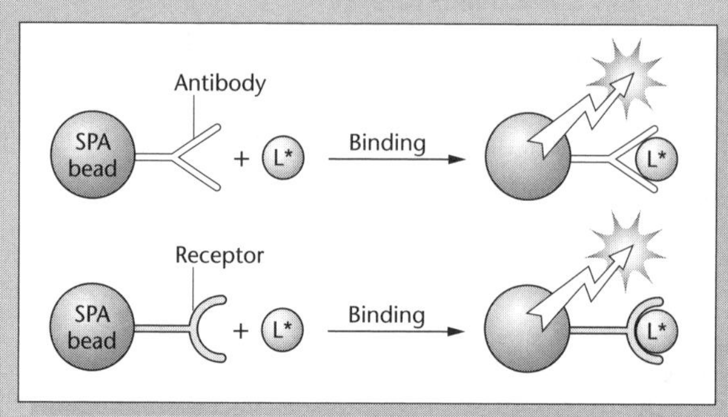

(2) The kinds of enzyme assays that have been converted to the SPA format include nucleic acid polymerases (e.g. reverse transcriptase), proteases (e.g. HIV protease), and transferases (e.g. farnesyl transferase). As with the ligand binding assays, the action of a polymerase or a transferase results in an increase in light output. By contrast, in the protease assay the starting reagents emit light and activity results in a decrease in light output.

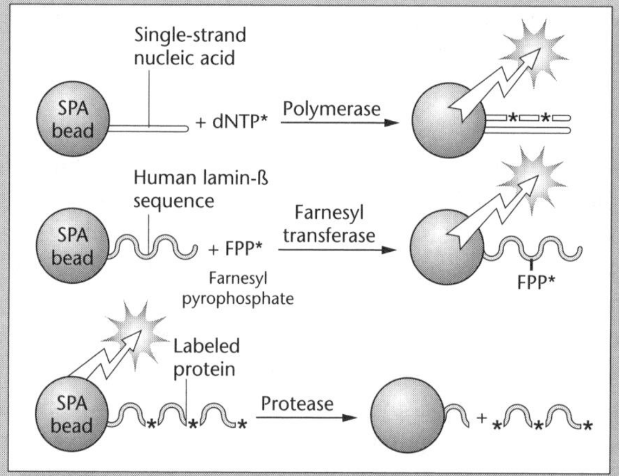

(3) Other enzymes that can be assayed using the SPA format include nucleic acid processing enzymes such as helicase and integrase.

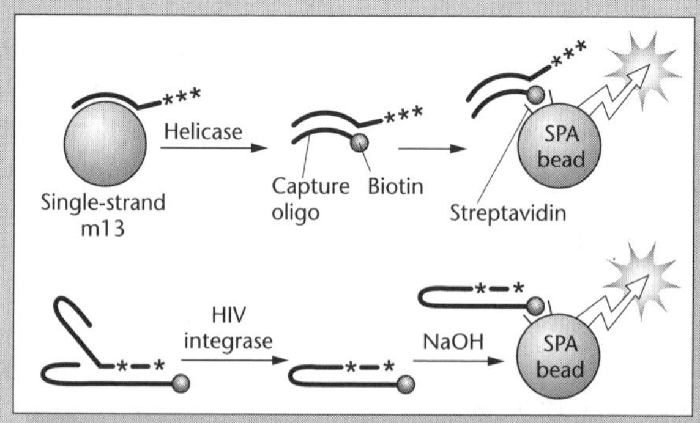

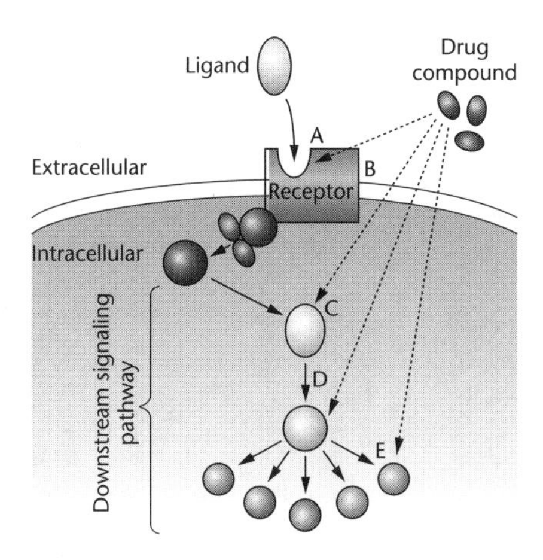

Fig. 7.4 Functional cell-based assays interrogate multiple potential points of drug interaction. A signal transduction pathway is portrayed, leading from a ligand–receptor interaction to downstream signaling events. Potential points of drug action are labeled A–E. Point A represents interaction of a drug with the same binding site as the ligand that might be present in a ligand-binding assay. B represents multiple alternative target sites that a drug might interact with to alter target function without competing for binding with a specific ligand in a ligand-binding assay. C–E represent other factors in the signal transduction cascade that a drug might interact with to alter the downstream response to receptor modulation. A functional cell-based assay may respond to drugs acting on A–E, whereas a ligand-binding assay is limited to interactions at A.

alone. This is achieved by first cloning essential genes (Table 7.2) under the control of a highly regulated promoter on a plasmid. The corresponding chromosomal gene then is removed by recombination using linear DNA transformation. This leaves a bacterial cell in which the intracellular level of one essential target protein can be adjusted with a specific inducer. The concentration of the target protein in the cell is varied and the strain then is screened against a chemical library using growth inhibition as the end-point. There are a number of advantages associated with this assay format. First, the strategy can be used with any target including proteins that are difficult to assay *in vitro*. Second, compounds identified in the screens are guaranteed to have antibacterial activity. Third, the screen for each different target is run with the same format and conditions thereby facilitating high-throughput operation.

Target validation and animal models

Prior to the advent of genomics, only a limited number of targets were available for each disease state and the identification of a new target by a pharmaceutical company would give that company a competitive edge. Today, using the techniques described in Chapters 3 and 4, the number of potential targets far exceeds the capacity of the pharmaceutical industry to put these targets into screens. Before limited resources are expended, poor targets must be eliminated through rigorous

Table 7.1 A comparison of cell-based uHTS systems.

Assay	Target classes	Cellular response	Measurement method	Specialized assay components	Time to detect response	Advantages	Disadvantages
R-SAT	GPCRs, nuclear receptors, kinases, and cytokine receptors	Receptor selection and amplification and cellular proliferation	Broad range of marker genes	Means to detect reporter	Days	Simple protocol, uHTS capacity, works with a variety of targets, target multiplexing	Response is distal from receptor modulation
Melanophore technology	GPCRs and cytokine receptors	Pigment dispersal	Light transmittance	Cultured melanophores, conditioned media	Seconds to minutes	Rapid response, potential to target multiplexing	Nonmammalian cellular background
FLIPR	GPCRs and ion channels	Calcium and voltage	Fluorescence – intensity	Integrated fluorescence plate reader and fluorescent probes	Seconds to minutes	Rapid proximal response	Dye loading, transient response may limit throughput
FRET-based voltage sensor	Ion channels	Fast voltage	Fluorescence – FRET	Fast voltage reader and fluorescent probe	Milliseconds to seconds	Ability to measure rapid membrane potential changes	Background signal, proprietary instrument requirement
Aequorin	GPCRs	Calcium	Luminescence	Coelenterazine	Seconds	Subcellular targeting	Flash luminescence, slow regeneration of probe
Transcriptional reporters	GPCRs, nuclear receptors, cytokine receptors, etc.	Transcriptional regulation	Broad range of marker genes	Means to detect reporter	Hours to days	Enzymatic amplification, variety of signaling inputs	Length of incubation, indirect response
cAMP immunoassay	GPCRs	cAMP	cAMP immunodetection	cAMP antibody	Seconds to minutes	Proximal measurement of second messenger	Complex protocol
FRET-based genetic sensors	Proteases and GPCRs	Calcium and protein cleavage	Fluorescence – FRET	Genetic GFP sensor constructs	Seconds to minutes	Noninvasive, allows subcellular targeting and ratiometric measurement	No enzymatic amplification of signal

GFP, green fluorescent protein; GPCRs, G-protein-coupled receptors; FLIPR, fluorometric imaging plate reader; FRET, fluorescence resonance energy transfer; R-SAT, receptor selection and amplification technology; uHTS, ultra-high throughput screening.

Table 7.2 Gene targets for cell-based antibiotic screening.

Gene	Gene product	Gene function
dnaB	Helicase	DNA replication
fabI	Enoyl-ACP reductase	Fatty acid biosynthesis
folA	Dihydrofolate reductase	Intermediary metabolism
gyrB	DNA gyrase B subunit	DNA replication
metG	Methionyl-tRNA synthetase	Protein synthesis
murA	UDP-acetyl-D-glucosamine enolpyruvyl transferase	Cell wall biosynthesis
pyrH	UMP kinase	Intermediary metabolism
tufA	Elongation factor Tu	Protein synthesis

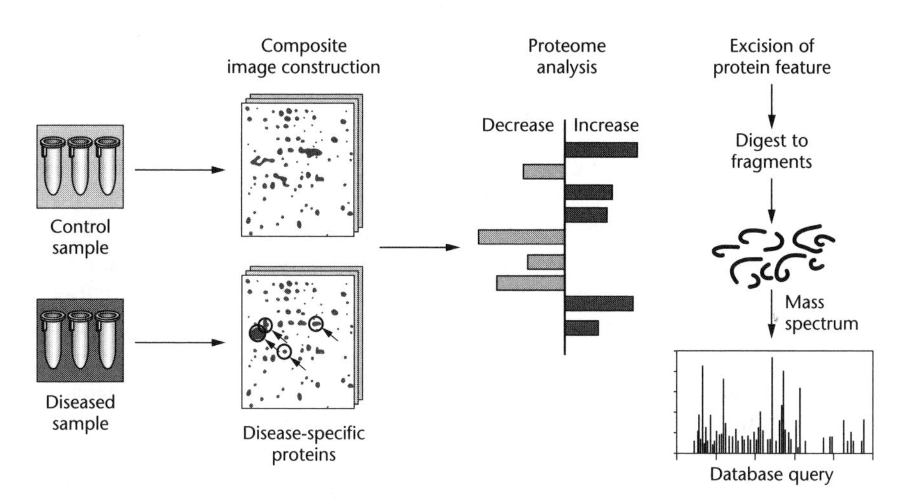

Fig. 7.5 Summary of the different stages of the proteomic process. Protein samples from control and disease groups are run on two-dimensional gels. Multiple images are generated and the resulting information is assembled into a proteomic database. The arrows indicate proteins that are specifically altered in disease samples. Protein profiles can be compared using appropriate software. The columns represent proteins that are increased or decreased on the gels. Protein features of interest are identified by excision from the gel, followed by protease digestion and MS analysis.

biologic assessment. Only those targets that demonstrate specific biology with therapeutic potential should be pursued otherwise poorly characterized targets could clog drug discovery pipelines.

Many different methods are being used to validate drug targets. These methods include bioinformatic categorization of a gene, analysis of expression levels in healthy versus diseased tissues (Fig. 7.5), cell-based assays, *in vivo* validation in an animal model and, the ultimate validation, testing new chemical entities in human clinical trials. The data produced by these different methods vary widely in their value to the drug discovery process. One method that generates high-quality

Box 7.2 Examples of drug targets validated in knockout mice models

Cathepsin K is an osteoclast-specific protease that cleaves bone matrix proteins such as Type I and II collagen, osteonectin, and osteopontin. Because of these enzymic activities it was felt that cathepsin K might have an important role in bone degradation and that an inhibitor of the enzyme might be useful in the treatment of osteoporosis. Knockout mice with a defect in the cathepsin K gene have impaired bone resorption, thereby confirming the role of the protease in bone matrix degradation and its relevance as a potential osteoporosis drug target.

Mice with disruptions of the melanocortin-3 and -4 receptor genes exhibit obesity and hyperinsulinemia. Clearly, these receptors play a key role in energy homeostasis and their activation could provide a treatment for obesity. Proof of this is provided by the observation that melanotan II, an agonist of the melanocortin receptor, increases metabolic rate and acts as an appetite suppressant in wild-type but not in knockout mice. An alternative way of treating obesity is suggested by the phenotype of mice with a knockout in the acetyl-CoA carboxylase 2 gene. These mutant mice eat more food than their wild-type littermates but burn more fat and store less fat in adipocytes.

information is the use of mouse models of human disease. Not only can these models be used for target validation but they also can be used to evaluate drug candidates. In general, these mouse models are generated by introducing mutations into genes of interest and hence are referred to as **knockout mice**. The kinds of genetic alteration that can be produced include deletions to disrupt gene function, conditional alleles that allow for tissue-specific or temporal control of gene expression, and point mutations. The techniques for generating these mutant mice are described later (see p. 197). Suffice it to say here that although the methodology can be lengthy, good progress is being made in industrializing the process.

Proof of the validity of using knockout mouse models is provided by the phenotypes of mice deficient for specific pharmaceutical target proteins that resemble the human clinical phenotype caused by the corresponding antagonist drug. For example, mice lacking the angiotensin-converting enzyme (ACE) or cycloxygenase-1 (COX1) have phenotypes resembling the effects on humans of ACE inhibitors (antihypertensive drugs) and nonsteroidal anti-inflammatory drugs. Examples of drug targets that are being actively pursued by the pharmaceutical industry and that have been validated using mouse models early in the drug development process are cathepsin K, the melanocortin-3 and -4 receptors, and acetyl-CoA carboxylase 2 (Box 7.2).

As noted above, animal models can be used to test the efficacy of new drug candidates as well as being used for target validation. One interesting application of animal models is in the search for conventional drug-based therapies to treat genetic diseases caused by trinucleotide expansions (see p. 91) and for which no treatment is available currently. A classic example of such a condition is Huntington's disease (HD). This is an autosomal-dominant neurodegenerative disorder caused by a CAG repeat expansion in the *HD* gene. The expanded repeats are translated into an abnormally long polyglutamine tract close to the N-terminus of the *HD* gene product (huntingtin). Studies in mouse models and humans suggest that the mutation is associated with a deleterious gain of function. There now is a wide range of mouse models for HD and these are providing powerful tools for preclinical testing of therapeutic strategies. Furthermore, HD is an adult-onset disease and so compounds can be tested to determine their ability to prevent HD onset as well as to treat the disease itself. So far, promising results have been obtained with minocycline, an antibiotic that is known to inhibit the transcription of caspases and nitric oxide synthase.

Combinatorial chemistry

Traditionally, chemical leads for the drug industry were identified through rational design and/or mass screening. When high-throughput screening was introduced it permitted the evaluation of hundreds of thousands of individual test molecules per year against a large number of targets. The new limitation was sourcing large libraries of chemicals. Until then, the compounds used in mass screening either consisted of a historical collection of synthesized compounds owned by pharmaceutical companies or natural product collections. Each of these has limitations. Historical collections contain a limited number of diverse structures (e.g. thousands of steroids, beta-lactams, etc.) and, while useful, represent only a small fraction of diversity possibilities. Natural products are limited by the structural complexity of the leads identified and the difficulty of reducing them to useful pharmaceutical agents. These limitations have been overcome by the use of combinatorial chemistry and combinatorial biosynthesis (see p. 192).

In traditional medicinal chemistry, drug candidates were synthesized one at a time and this was a time-consuming and labor-intensive process. In combinatorial chemistry, large numbers of compounds (called libraries) are synthesized by performing reactions in a manner that produces large combinations of products. These libraries often are mixtures. This approach has been called **irrational drug design** since it involves making a large pot of all possible chemical combinations of several reactants. A related technique is **multiple parallel synthesis** where the same reactions are repeated separately to produce many individual but related products. That is, a compound library is constructed by synthesizing many compounds in parallel with each compound being kept in a separate reaction vessel.

Combinatorial chemistry is best illustrated by reference to the chemical synthesis of peptides. In the classic Merrifield method, an amino acid is attached to a solid support such that the amine group is free. This immobilized amino acid is reacted with an N-protected amino acid to generate a dipeptide. After deprotecting the amine group this dipeptide can be reacted with another N-protected amino acid to generate a tripeptide. This cycle can be repeated many more times. If instead of adding one N-protected amino acid at each cycle we added 20 different amino acids we would generate 20 dipeptides, 400 tripeptides, 8000 tetrapeptides, and so on. As Fig. 7.6 shows, these peptides could be synthesized as a mixture or individually. It should be apparent from Fig. 7.6 that if all the peptides are synthesized as a mixture then identifying all the components will be very difficult. On the other hand, if the peptides are

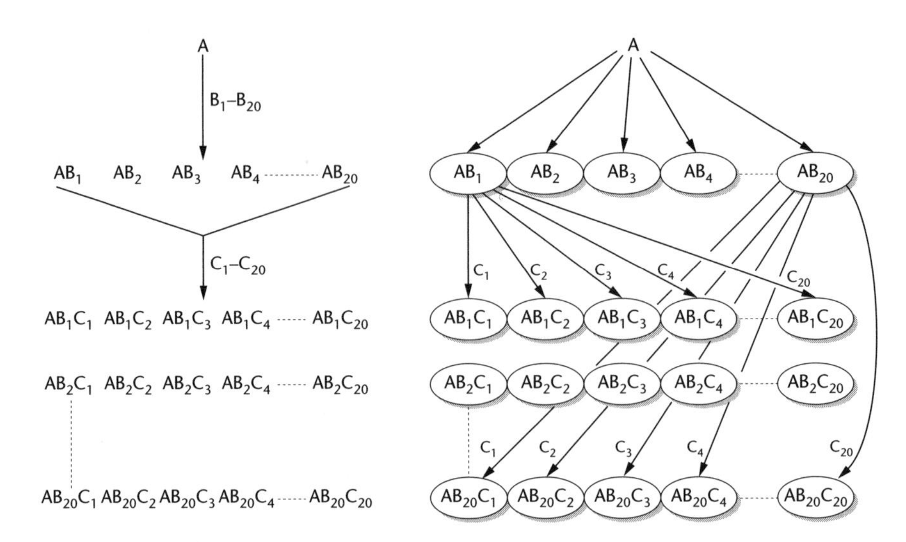

Fig. 7.6 The two different ways of synthesizing random sequence peptides. For clarity, the process has been shown only as far as a tripeptide.

synthesized separately the number of separate reactions soon becomes very large, e.g. 8000 tetrapeptides will be converted to 160,000 pentapeptides. Another problem is that at each cycle the different reactants may combine at different rates and with different yields and this can result in vastly different amounts of the final products. These problems can be overcome by using the **split synthesis** method (Fig. 7.7).

An important component of the split synthesis method is deconvolution of the library that is formed. An example of deconvolution is shown in Fig. 7.8 for a molecule with four variable substituents (A–D). If there are five different possibilities for each substituent then the total library consists of $5^4 = 625$ compounds. In the first deconvolution, 25 sublibraries are synthesized, each containing 25 compounds where the substituents A and B are defined. These sublibraries are tested in the selected screening system and the most active sublibrary identified (A^2B^1 in Fig. 7.8). Based on this, five further sublibraries in the form $A^2B^1C^nD^{1-5}$ are prepared, each containing five components in which C now is defined. The optimal residue D is determined in the last step by synthesis of the five individual compounds in the form $A^2B^1C^3D^n$. In this example the most active compound is $A^2B^1C^3D^5$.

The techniques described above were developed originally for the generation of peptide libraries. However, peptides generally do not make good drug candidates because of their *in vivo* instability and poor oral absorption. For this reason library generation is now focused on the small organic molecules favored by drug companies. The first success of this approach was the generation of a library of substituted benzodiazepines and since then many others have been described.

In theory, any chemical template can act as the starting material and every possible functional group can be added at each position in the template. In practice, medicinal chemists tend to choose templates and functional groups based on prior knowledge of bioreactive compounds. Analysis of the structures of known drugs is

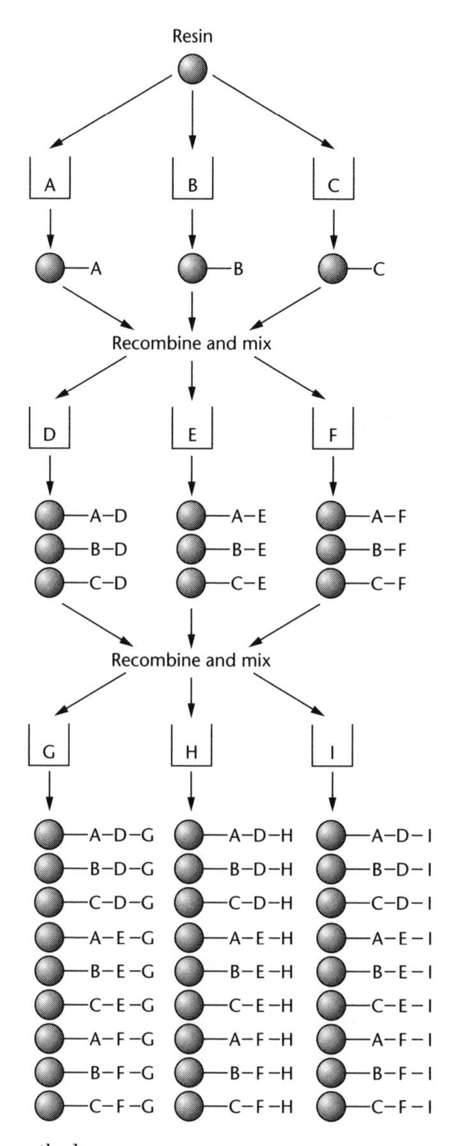

Fig. 7.7 Split synthesis method.

used to provide the initial design and building blocks for creating plausible active compounds. Combinatorial synthesis is used to try and emulate or improve upon this knowledge. There are two reasons for this more conservative approach. The first of these is the Lipinski "rule of 5" (Table 7.3). Analysis of existing drugs has shown that less than 80% fall below the cut off for each criterion. The second reason is that the majority of drugs are formed from only about 30 geometric frameworks (using atoms as vertices and bonds as edges). Such frameworks make an obvious start or a potential goal for combinatorial synthesis.

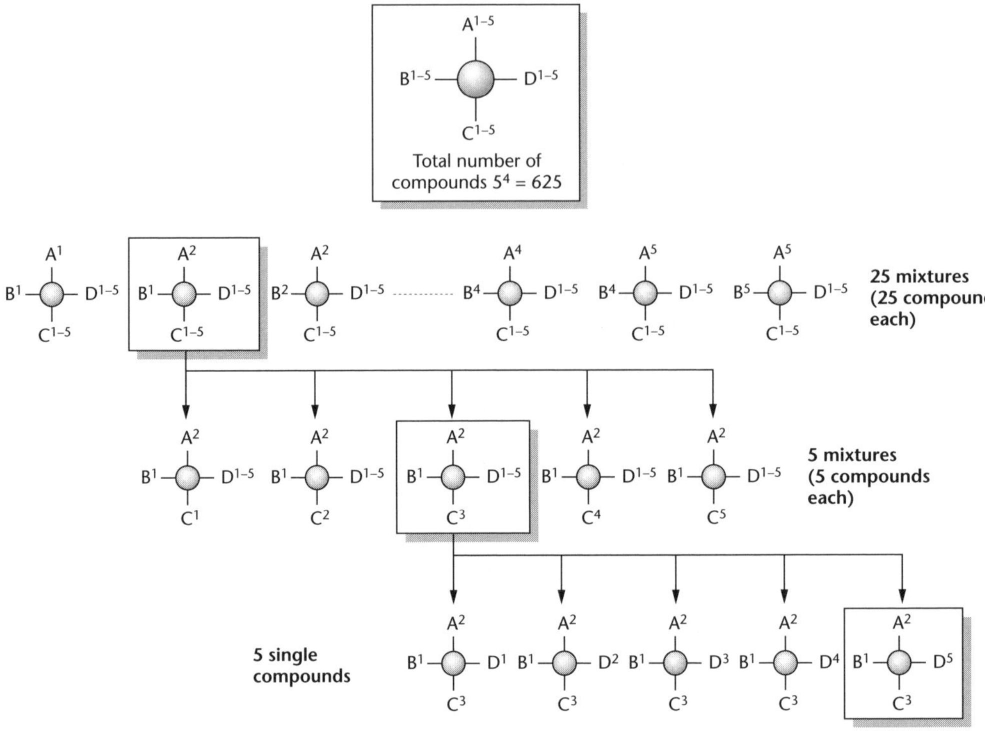

Fig. 7.8 Identification of the most active library component by deconvolution.

Table 7.3 Structural characteristics of drug candidates that can lead to poor adsorption and permeability

- There are more than five hydrogen-bond donors (expressed as the sum of -OHs and -NHs)
- The molecular weight is more than 500
- The logP is more than 5 (Log P is the logarithm value of the partition coefficient)
- There are more than 10 hydrogen-bond acceptors (expressed as the sum of nitrogens and oxygens)

Dynamic combinatorial libraries

Although combinatorial techniques have allowed the synthesis of very large arrays of compounds in a short time, each individual compound needs to be prepared and then characterized. If, however, the target substance itself could be used to select an active molecule directly from the library pool then the screening process would be more efficient and greatly simplified. In addition, if the library pool itself were able to undergo changes in composition during the process, so as to adapt to the target constraints, then the screening signal would be amplified thereby facilitating detection and characterization. Furthermore, if the active species could be analyzed directly while bound to the receptor site then several synthetic steps could be avoided. This approach is known as **dynamic combinatorial chemistry**.

For the efficient production of a dynamic combinatorial library (DCL) the building blocks need to fulfill several important characteristics. First, they must possess functional groups that can undergo reversible exchange. Second, they must cover as completely as possible the geometric and functional space of the potential target. Third, these recognition groups need to be organized geometrically for optimal binding to occur. In theory, library generation can be accomplished using any type of reversible physical or chemical mechanism so long as the respective interconverting states can be properly controlled and the final products identified. In reality, it is essential that the reaction mechanisms are compatible with biologic targets for it is the addition of the target that favors the formation of the best-binding constituent.

Three approaches to DCL generation and screening have been developed. These have a common first reversible generation step but differ in the screening and selection step. In the adaptive approach, the generation of the DCL is done in the presence of the target, resulting in amplification of the best-bound species; that is, screening takes place simultaneously with library generation. In the pre-equilibrated approach, generation of the library is achieved under reversible conditions but the identification and screening is done under static conditions. No amplification can take place in this case but the protocol is of value when the biologic target is very sensitive or not available in large amounts. In the iterative approach, the DCL is generated in one compartment under appropriate conditions and then allowed to interact with the target. Unbound species are retransferred to the reaction chamber, rescrambled, and again allowed to interact with the target. After several rounds of synthesis and screening the accumulated active species can be analyzed.

Virtual screening

In virtual or *in silico* screening the starting point is a three-dimensional structure of the target molecule. Potential ligand binding sites on this structure are identified and these are modeled using sophisticated software programs. A large number of molecules then are screened to identify those that have the correct geometric and electronic features to fit the designated sites. The molecular structures to be screened might be an existing compound collection or a virtual collection of molecular structures obtained from a preferred set of combinatorial chemistries. The advantage of the virtual set is that it is not necessary for the structures to have been synthesized before the *in silico* docking experiments.

The availability of a three-dimensional structure of a target is an essential prerequisite for virtual screening. Two techniques have been used to generate these structures: X-ray crystallography and nuclear magnetic resonance (NMR) spectroscopy. The former has proved a very versatile method, with many globular macromolecules proving to be crystallizable, and with no limitations on the size and complexity of the macromolecules or their assemblies. NMR has the advantage of being carried out in concentrated solutions rather than in crystals. Comparative studies using the two methods can identify places where crystal contacts disturb the local structure. Although NMR can define certain dynamic properties of macromolecules it is effectively limited to those molecules with molecular weights less than 30 kDa. For this reason, X-ray crystallography is the most widely used method of structure determination. Until recently, the biggest limitation in X-ray crystallography was the effort required to generate good quality crystals. Now, many of the steps have been automated and as a result the structures of over 5000 different wild-type proteins have been deposited in the Protein Data Bank.

Combinatorial biosynthesis and chemobiosynthesis

A surprising number of drugs in use today are derived from plants and microorganisms. Most of these natural products are characterized by complex structures (Fig. 7.9) with many chiral centers, and selective chemical modification to generate new molecules is very difficult. This is unfortunate since they have many attributes that make them good drug candidates. However, now that the genetics of the biosynthesis of these molecules is understood, a generic method for synthesizing new variants has been developed. This is known as combinatorial synthesis and the principle is shown in Fig. 7.10. The best example of combinatorial synthesis is the development of novel polyketides.

Polyketides have little structural resemblance to each other (Fig. 7.9). They get their name because they share biosynthetic pathways that produce a common ketone structure among their synthesis intermediates. All polyketides are built from a linear polymer of carbon atoms created by sequential reactions of enzyme complexes called polyketide synthases (PKSs). Anywhere from 5 to 50 PKS reactions are required, depending on the complexity of the end molecule, and the process is analogous to the synthesis of fatty acids. Once synthesis of the linear chain is complete, the chain is released from the PKS complex. It then is cyclized by non-PKS enzymes and subjected to further modification such as methylation, hydroxylation, addition of sugars, etc.

A PKS is not a single protein. Rather, it is composed of several polypeptides having, at a minimum, loading, chain-extension, and chain-releasing activities. Between extension and untethering, a PKS sometimes sandwiches in one or more ketone-modifying reactions. Extension and modifier polypeptides are together called modules and a PKS may contain one module or several. Because the genes for all the modules are linked it is very easy to isolate them and to clone them in new hosts such as *Escherichia coli* and yeast. Once the gene clusters have been isolated from several different organisms it is possible to undertake gene shuffling of the kind shown in Fig. 7.10 and thereby generate novel molecules. For example, by mixing genes for the biosynthesis of actinorhodin with those for granaticin and medermycin it was possible to generate two new antibiotics mederrhodin A and

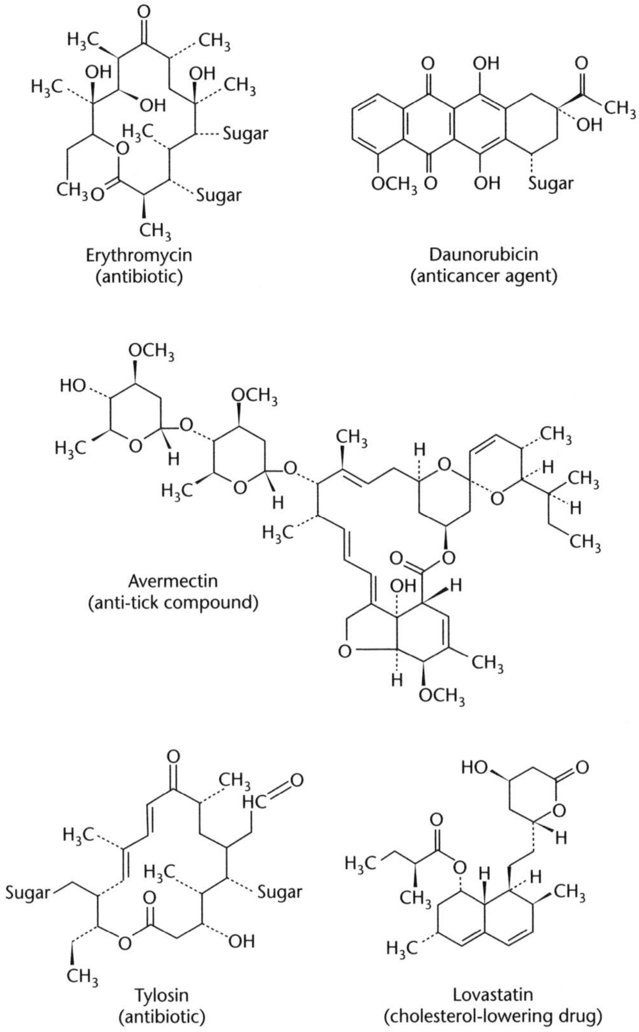

Erythromycin
(antibiotic)
Daunorubicin
(anticancer agent)
Avermectin
(anti-tick compound)
Tylosin
(antibiotic)
Lovastatin
(cholesterol-lowering drug)

Fig. 7.9 Some examples of polyketides.

Antibiotic A
Antibiotic B
Antibiotic C
New antibiotic X
New antibiotic Y

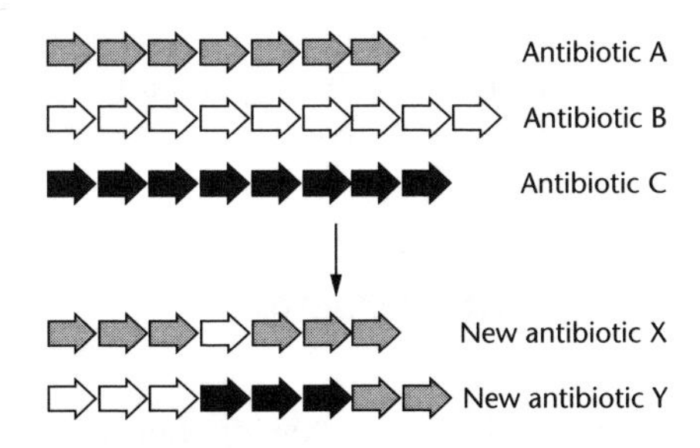

Fig. 7.10 The creation of new antibiotics by shuffling genes for antibiotic synthesis.

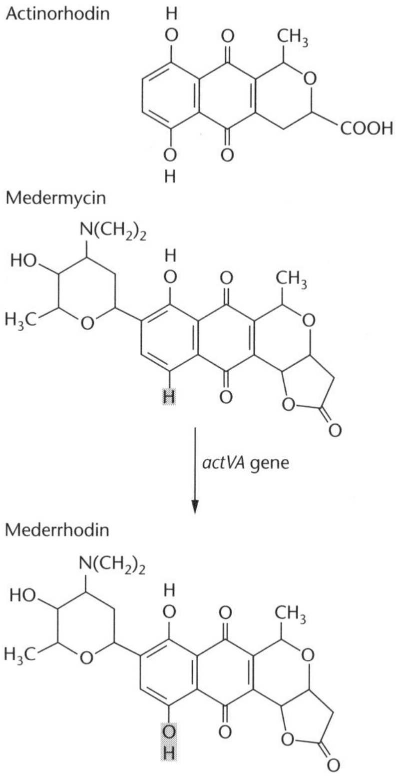

Actinorhodin

Medermycin

actVA gene

Mederrhodin

Fig. 7.11 The formation of the new antibiotic mederrhodin from medermycin by the actVA gene product.

dihydrogranatirhodin (Fig. 7.11). The process can be taken a step further in that the cloned genes can be subjected to random mutagenesis and then screened to determine if the PKSs have new activities.

Another way of generating novel polyketides is to overcome the preference of the PKS enzymes for processing compounds produced by the host. This is done by altering a gene so that the extension step in the first module is blocked. This prevents production of second step precursors and effectively blocks natural biosynthesis. The cells then are fed synthetic precursors and those precursors that can act as substrates are converted by the PKS enzymes into novel metabolites. This procedure is known as **chemobiosynthesis**.

Drug metabolism

Most drugs are metabolized by the body following their administration and there are two important facets of this metabolism. The first of these is the rate of metabolism. If a drug is metabolized too fast it may not be effective. On the other hand, if it is metabolized too slowly it may accumulate in the body with each successive dose thereby leading to toxicity. This aspect of metabolism is controlled by the genetic make-up of the drug recipient and was described in the section on pharmacogenomics (see p. 108) in Chapter 4. The other important aspect is the identity of the metabolites that are produced following administration. If these metabolites are known to be toxic to humans, or share molecular characteristics with known toxicants, then further development of the drug is likely to be halted.

The standard method of determining the metabolic fate of a new drug is to administer a radiolabeled version of it to selected animal species and then to analyze all the labeled compounds in the urine and feces using a variety of sophisticated analytical techniques. In these studies it is important to use animal species whose metabolism is believed to be most like that of humans. The significance of this is illustrated by the sequencing of the mouse genome (see Box 4.2), which showed that compared with humans mice have more genes for the metabolism of xenobiotics. Thus a mouse may metabolize a drug in a different way to humans. The genomes of many of the other animals used in drug evaluation are being sequenced and once this task is complete we will know which species really are the best test candidates.

The methodology described above is both difficult and costly and it would be preferable if some kind of prescreening could be undertaken. One way of doing this prescreening is to use recombinant cells carrying genes for human enzymes involved in drug metabolism, e.g. cytochromes P450. The genes for these cytochromes have been cloned in *E. coli*, yeast and animal cells and make metabolic studies very easy. The recombinant cells are grown in culture, the drug is added, and after a suitable incubation period the metabolites are detected by mass spectrometry without the need to extract them from difficult matrices such as feces. If metabolites of known or suspect toxicity are identified then progression to studies in live animals is unlikely.

Toxicogenomics

The human safety of new drugs is assessed by feeding the drugs to animals and then determining if any adverse anatomic, histologic or biochemical changes have occurred. The process is laborious and costly and is not 100% predictive since some promising therapeutics are withdrawn from the marketplace because of unforeseen human toxicity. It also is the stage in drug development where most bottlenecks occur and so the availability of high-throughput techniques, or ones more predictive of human responses, would be most welcome. One such technique is **toxicogenomics**: the examination of changes in gene expression following exposure to a drug candidate. Toxicogenomics offers the potential to identify a human toxicant earlier in the drug development process and to detect human-specific toxicants that cause no adverse reaction in rats.

There are two fundamental ways of collecting and using gene expression data to predict toxicity. In the first of these, one identifies those genes that are expressed

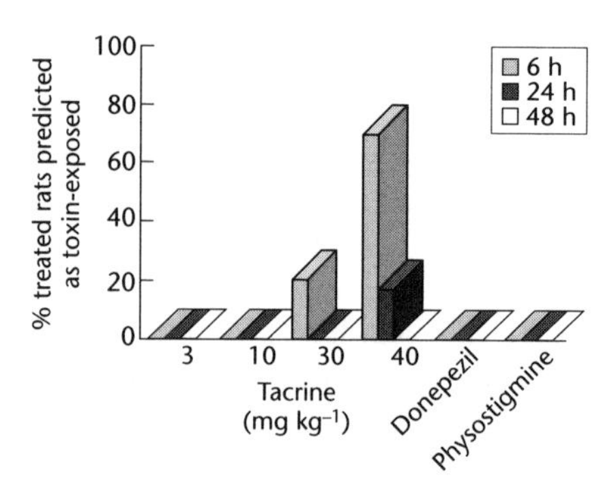

Fig. 7.12 Animals were dosed orally once with tacrine, donepezil, or physostigmine, and then killed 6, 24, or 48 h later. The percentage of animals identified, on the basis of modeling software, as exposed to toxicants or nontoxicants is illustrated.

following administration of the test drug and compares the result with that obtained by administering one or more other drugs of the same class whose toxicity to humans is known. In the second approach, a large database is constructed with expression data for marketed pharmaceuticals, classic chemical toxicants and proprietary drugs and that spans a wide range of pharmacologic and structural classes. These data are analyzed and used to generate algorithms that will predict toxicity of a new compound. Whichever method is used, care must be taken to distinguish pharmacologic effects from toxicologic ones by using multiple doses of each compound.

As an example of the use of toxicogenomics consider three drugs (tacrine, donepezil, and physostigmine) that have been developed to treat Alzheimer's disease by the inhibition of acetylcholinesterase. During preclinical development there was no evidence of liver toxicity in mice, rats or dogs for any of the three drugs. However, 25% of humans treated with tacrine exhibited asymptomatic elevation of serum aminotransferase and biopsy revealed liver necrosis. For this reason, tacrine is viewed as a human-specific toxicant. The data in Fig. 7.12 show that toxicogenomic algorithms correctly predicted that animals treated with high doses of tacrine had been exposed to a toxicant. Animals treated with donezipil, physostigmine or low doses of tacrine were designated as toxin-exposed.

The gene expression studies that underpin toxicogenomics are undertaken using microarrays. Initially, the selection of genes that was used on the microarrays was those available on commercially available chips. However, as data have accumulated on a wide range of drugs particular genes and families of genes are being recognized as being linked to toxicity. The ultimate goal is to have microarrays carrying only those genes whose expression changes on administration of a toxicant.

Further reading

POGM: Chapter 5 details the methods for placing genes under the control of regulatable promoters and Chapter 11 describes how to introduce targeted mutations into laboratory animals.

POGA: Chapter 9 describes in detail different methods of monitoring gene expression and supplements the material covered in this chapter.

Abuin A, Holt KH, Platt KA *et al.* (2002) Full-speed mammalian genetics: *in vivo* target validation in the drug discovery process. *Trends Biotechnol* **20**, 36–42.

Bandara LR, Kennedy S (2002) Toxicoproteomics – a new preclinical tool. *Drug Discovery Today* 7, 411–418.

Castle AL, Carver MP, Mendrick DL (2002) Toxicogenomics: a new revolution in drug safety. *Drug Discovery Today* 7, 728–736.

These two papers give complementary insights into the science of toxicogenomics.

Battersby BJ, Trau M (2002) Novel miniaturized systems in high-throughput screening. *Trends Biotechnol* **20**, 167–173.

Blundell TL, Jhoti H, Abell C (2002) High-throughput crystallography for lead discovery in drug design. *Nature Rev Drug Discovery* **1**, 45–54.

An excellent review from a leading authority in the field.

Croston GE (2002) Functional cell-based uHTS in chemical genomic drug discovery. *Trends Biotechnol* **20**, 110–115.

Drews J (2000) Drug discovery: a historical perspective. *Science* **287**, 1960–1964.

This is an excellent overview from one of the most respected figures in pharmaceutical R&D. The same issue of *Science* has a collection of other articles on genomics and drug discovery.

Goodnow RA (2001) Current practices in generation of small molecule new leads. *J Cell Biochem* **37** (Suppl), 13–21.

Knowles J, Gromo G (2003) A guide to drug discovery: target selection in drug discovery. *Nature Rev Drug Discovery* **2**, 1–10.

Pfeifer BA, Khosla C (2001) Biosynthesis of polyketides in heterologous hosts. *Microbiol Mol Biol Rev* **65**, 106–118.

This review provides the background to combinatorial biosynthesis and chemobiosynthesis.

Ramstrom O, Lehn J-M (2002) Drug discovery by dynamic combinatorial libraries. *Nature Rev Drug Discovery* **1**, 26–36.

Rubinsztein DC (2002) Lessons from animal models of Huntington's disease. *Trends Genet* **18**, 202–209.

CHAPTER EIGHT

Gene and cell therapies

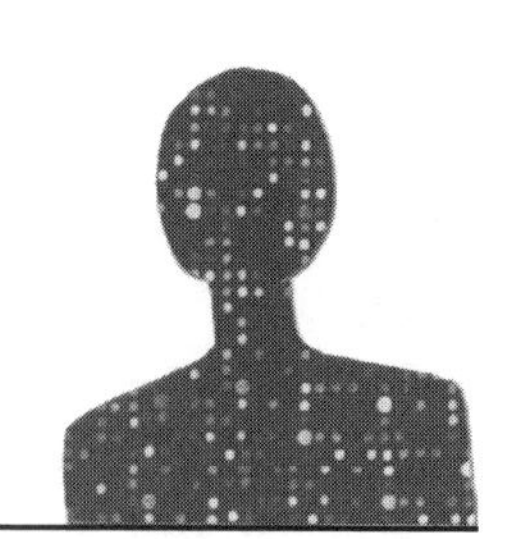

Introduction

Most conventional drugs are either proteins or small molecules that interact with proteins. In other words, they act at the protein level rather than at the level of the underlying genes. For infectious diseases and cancer, conventional drug treatments can in the best cases be expected to effect a complete cure, i.e. the pathogen or the group of abnormal cells is eradicated. For inherited diseases, however, drugs can only treat **symptoms** while the underlying gene defect remains unchanged. An alternative approach in such cases is **gene therapy**: the use of nucleic acids in an attempt to repair the malfunctioning DNA sequence or at least introduce a compensatory change that will restore the normal physiologic functions of the cell. Gene therapy is one component of an emerging group of therapies collectively described as **gene medicine**, in which nucleic acids are introduced into human cells. Another form of gene medicine is the use of nucleic acids in the same manner as conventional drugs, though the target of the therapy is often the mRNA produced by the malfunctioning gene rather than the protein. Also, the use of **DNA vaccines**, which express antigens in the body, comes under the heading of gene medicine.

Gene therapy is, in essence, the opposite approach to the creation of animal **disease models** by mutagenesis. In this application of gene transfer technology, mutations that resemble human diseases are deliberately engineered in animals, allowing the biochemical and physiologic aspects of the disease to be investigated and novel drugs to be tested for efficacy. Many of the techniques employed in gene therapy and disease modeling are based on exactly the same principles, although with different aims. Further disease models can be produced using nucleic acids or proteins to interfere with gene expression or protein activity.

Gene transfer technology is becoming increasingly integrated with **cell therapy**, the use of whole cells derived either from the patient or an alternative source as a therapeutic modality. While representing perhaps the most promising of therapeutic strategies for long-term illnesses, both gene therapy and cell therapy raise important ethical and safety issues that need to be addressed.

Box 8.1 Ethical concerns raised by gene therapy

Both somatic and germ-line gene therapy raise a number of significant ethical concerns. One common issue is the **potential for abuse**. Gene therapy could be used not only to cure devastating diseases (or eliminate them from a family in the case of germ-line therapy) but also to enhance subjectively defined favorable characteristics and suppress unfavorable ones. On a small scale, this would result in a generation of **designer children**, with traits chosen by their parents. On a large scale, gene therapy could be used to manipulate the genetic constitution of entire populations (**eugenics**). There are strong moral arguments that favor the use of gene therapy to cure diseases that cannot be treated in any other way, but an important issue is where the diseases stop and the nondisease characteristics begin. If gene therapy could be used to raise the IQ of a child from 40 to 100, would this be a disease treatment or a genetic enhancement? And who gets to choose what is acceptable and nonacceptable use of gene therapy?

Germ-line gene therapy has two other major ethical implications:

1 **Imperfect technology**. The effects of stable gene transfer are unpredictable, so even if the target disease was cured, further defects could be unknowingly introduced. The disastrous outcome of the recent somatic gene therapy trials to cure SCID provides a clear example of the potential dangers that could arise in germ-line gene therapy (see main text).

2 **No right to decide**. Individuals resulting from germ-line gene therapy would be unable to give informed consent for the procedure and would therefore have no say in whether their genetic material should be modified. This constitutes a denial of human rights.

Gene therapy

Gene therapy is defined as a therapeutic strategy in which a patient's cells are genetically modified in an attempt to alleviate or cure a disease. There is an important distinction between **somatic gene therapy**, where modifications are introduced into somatic cells and are confined to the patient, and **germ-line gene therapy**, where modifications are introduced into the cells that give rise to gametes – and can therefore be passed to subsequent generations. Only somatic gene therapy is currently permitted. While there are compelling reasons to allow germ-line gene therapy under some limited circumstances (e.g. where a parent is certain to pass on a disease-causing mutation to his or her offspring), the procedure has been banned universally because of its ethical implications (Box 8.1).

Several different types of somatic gene therapy can be envisaged, and the most suitable approach depends on the nature of the disease (Fig. 8.1):

- **Gene augmentation therapy**. This is appropriate for the treatment of inherited disorders caused by the loss of a functional gene product. The aim is to add a functioning copy of the lost or mutated gene back into the genome and express it at a level sufficient to replace the missing protein. It is only suitable if the effects of the disease can be reversed.
- **Gene inhibition therapy**. This is suitable for the treatment of infectious diseases, cancer and inherited disorders caused by inappropriate gene activity (gain of gene function). The aim is to introduce a new gene whose product inhibits the expression of the pathogenic or malfunctioning gene, or interferes with the activity of its product.
- **Gene replacement therapy**. This is suitable for inherited disorders only (not infectious diseases or cancer) but can be used to treat diseases characterized by

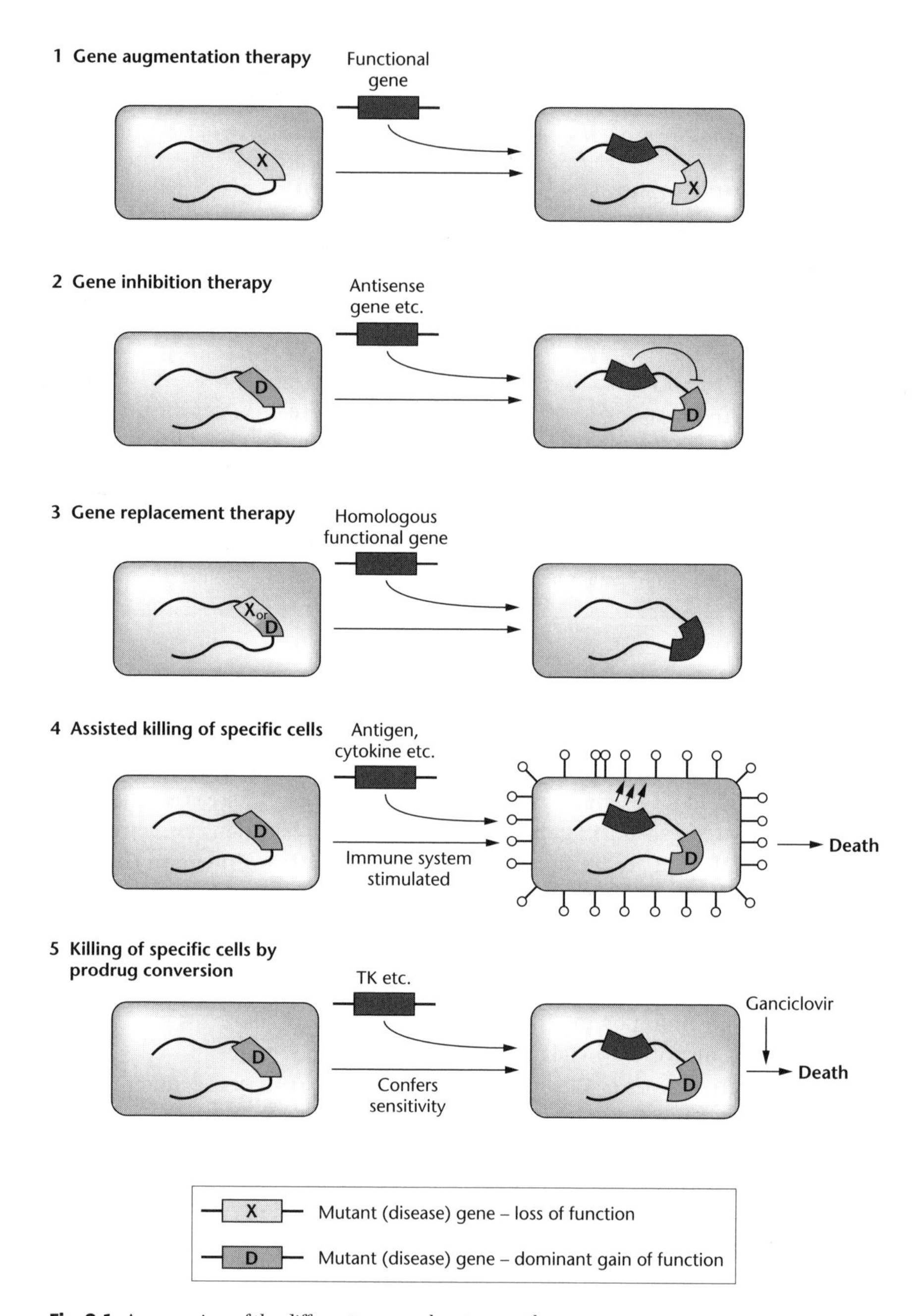

Fig. 8.1 An overview of the different approaches to gene therapy.

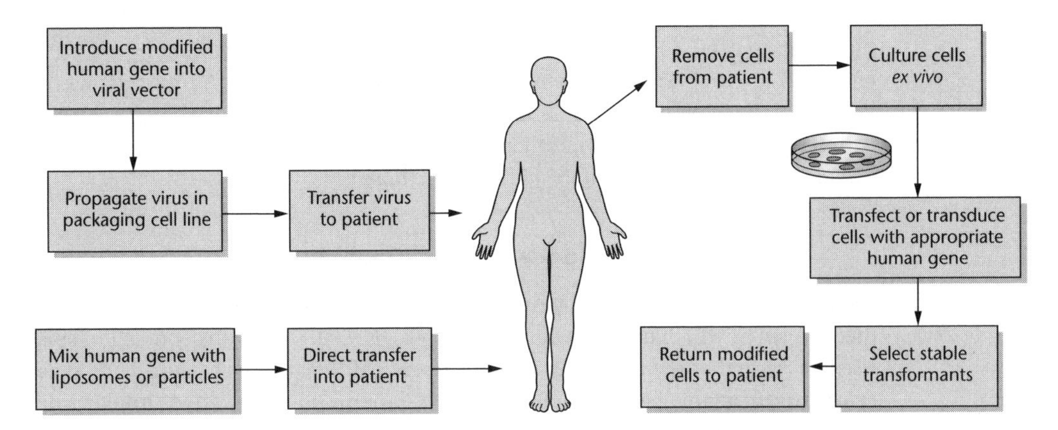

Fig. 8.2 Strategies for gene therapy. Left: *in vivo* gene therapy using viral and nonviral vectors. Right: *ex vivo* gene therapy using viral and nonviral vectors.

either loss or gain of function. The aim is to introduce a normal functioning copy of the nonfunctional target gene and then persuade the exogenous DNA to undergo recombination with the target and therefore replace it. While potentially the most straightforward way to correct genetic defects, this does rely on homologous recombination which is a very inefficient process in most cells.

- **Killing of specific cells**. This is suitable for diseases such as cancer that can be cured by eliminating certain populations of cells. The aim is to express within such cells a **suicide gene**, whose product is toxic. One approach is the expression of a protein that makes the cells vulnerable to attack by the immune system. Another involves the expression of an enzyme that converts a harmless substance (a prodrug) into a more toxic molecule. Owing to the lethal effects of suicide genes, they must be directed to the target cell types with great accuracy to avoid side effects.

Gene delivery strategies

Genetic diseases are usually manifest in specific populations of cells. Sometimes, as is the case for cancer, this is because the disease-causing mutation is present only in that cell population. For inherited disorders, the mutation is usually present in all cells but the **effects** are restricted to those cells where the gene is expressed (or where normal gene expression is lost). The method chosen for DNA delivery in gene therapy therefore depends to a large degree on the accessibility of the **relevant cells** (Fig. 8.2).

If cells can be removed and cultured without harming the patient, they can be genetically modified in culture and then reintroduced into the patient's body. This ***ex vivo* gene therapy** strategy is applicable for the treatment of blood and immune system disorders because hematopoietic stem cells can be removed, cultured, and transformed relatively easily. Furthermore, they persist for a long time when introduced back into the patient and can therefore give rise to an indefinite number of (genetically modified) daughter cells. *Ex vivo* gene therapy allows the target cells to be monitored carefully while in culture, so that the correct type of genetic modification can be selected. For inaccessible cells, or cells that cannot be cultured efficiently, gene therapy involves the direct introduction of DNA into cells while they are still in the body. This is known as ***in vivo* gene therapy**, and the gene delivery system must be both efficient and selective for particular target cell types. This is especially important if the aim is to kill those cells.

Gene delivery mechanisms

As well as the overall strategy for gene delivery, it is also necessary to consider the mechanism by which nucleic acids gain entry to the cell. Gene transfer mechanisms used for gene therapy are either **viral** or **nonviral**. Viral delivery, also known as **transduction**, involves the packaging of DNA (or in some cases RNA) into a virus particle. Gene transfer occurs by the normal viral infection route and is therefore both efficient and cell selective. For this reason, viral delivery is the preferred strategy for *in vivo* gene therapy. The different viruses currently used or under development as gene therapy vectors are compared in Box 8.2.

It is important to remember that viruses are pathogenic entities, so steps must be taken to disable them otherwise the vectors themselves could cause disease.

Box 8.2 Properties of viral vectors for gene delivery

Viral vectors used for gene therapy are evaluated on the basis of their efficiency, capacity for foreign DNA, safety, transformation mechanism (episomal or integration), persistence, and tropism. No one virus is suitable for all applications, though there is increasing interest in the design of hybrid viruses which incorporate the best features of different types of vector.

Adenoviruses

These are DNA viruses that cause benign infections of the upper respiratory tract, i.e. common colds. Transformation is episomal and the virus can infect a wide range of dividing and nondividing cells. Advantages: efficiency (high titers can be produced); capacity (up to 35 kb of foreign DNA). Disadvantages: persistence (expression is usually short-lived, although the latest vectors have a greater longevity and are particularly useful for long-term gene expression in neurons); safety (adenovirus vectors have been shown to provoke inflammatory responses in several patients). Such an inflammatory response resulted in the death of 18-year-old Jesse Gelsinger during a phase I gene therapy trial for the inherited liver disorder ornithine transcarbamylase (OTD) deficiency. All gene therapy trials involving adenovirus have been halted while safety procedures are examined. It emerged after the trial that the majority of severe responses to the virus had not been reported.

Adeno-associated viruses

Adeno-associated viruses (AAVs) are single-stranded DNA viruses that cause asymptomatic infections. Transformation is by stable integration and the virus can infect a wide range of dividing and nondividing cells. Advantages: safety (the virus requires the presence of adenovirus or herpes virus to replicate and is therefore naturally replication defective); persistence (stable integration allows long-term gene expression). Disadvantages: capacity (less than 5 kb of DNA can be accommodated). The wild-type virus has an interesting safety feature in that it tends to integrate at a defined site in the human genome on chromosome

19 thus reducing the risk of insertional mutagenesis (cf. retroviruses below). Unfortunately, this property is conferred by the viral Rep protein, but the *rep* gene must be deleted in order to provide room for foreign DNA.

Baculoviruses

Baculoviruses are DNA viruses that usually infect insects, but they can also infect human cells. Transformation is episomal and while initial reports suggested that liver cells were preferentially infected, it now appears the virus will infect a much wider range of cells. Advantages: safety (the virus infects human cells but cannot replicate in them); capacity (the virus has a rod-shaped capsid and therefore no defined upper limit for foreign DNA). Disadvantages: efficiency (virus is sensitive to complement-mediated inactivation, although strategies have been developed to overcome this). Unlike the other four types of virus discussed, baculoviruses have yet to be used in gene therapy trials.

Herpesviruses

Herpes simplex virus (HSV) is a DNA virus that displays extraordinary neurotropism, which makes it particularly attractive for neural gene therapy. Transformation is episomal and the virus can spread through the synaptic network. Advantages: capacity (a precise upper limit for foreign DNA has not been established); persistence (lifelong latent infections in neurons can be established).

Retroviruses

Retroviruses are RNA viruses that possess a reverse transcriptase, enabling them to synthesize a cDNA copy of their genome. This copy is integrated into the host genome, resulting in stable transformation and longlasting transgene expression. Typical retroviruses only infect dividing cells, which makes them particularly suitable for cancer gene therapy. However, vectors based on the lentiviruses (such as HIV) can infect nondividing cells. Advantages: efficiency (high titers can be produced and infection is very efficient); persistence (stable integration makes long-term gene expression possible). Disadvantages: capacity (can accept a maximum of approximately 8 kb of foreign sequence); safety (integration is random and there is a danger of insertional mutagenesis). The integration mechanism can result in the activation of adjacent genes, including oncogenes, which is exactly what appears to have occurred in the recent SCID gene therapy trials (see main text). Another safety concern involves the use of HIV for vector construction; this is the only virus used for gene therapy which is known to cause a lethal disease. Extreme safety precautions are therefore necessary to avoid replication-competent viruses being produced by recombination.

The usual method is to remove essential genes and therefore make the viruses **replication defective**, an approach which also increases their capacity for foreign DNA. In order to prepare large quantities of such crippled viruses, the missing functions must be supplied from an alternative source. This may mean using a **helper virus** to make the missing gene products but in most cases a **packaging line** is used, i.e. a cell line that is stably transformed with the appropriate viral genes (Fig. 8.3). Once the DNA is packaged, the defective virus can be isolated from the packaging line. It can then infect its target cell and introduce its cargo of DNA or RNA, but it cannot replicate and cause disease symptoms. Safety features must be built into packaging lines to discourage recombination between defective viruses and the genes supplying helper functions, otherwise replication-competent viruses could be produced.

Nonviral gene delivery methods are considered to be safer than viruses. They include **transfection** (where cells are persuaded to take up DNA from their surroundings) and **direct transfer** (where DNA is introduced into cells by physical

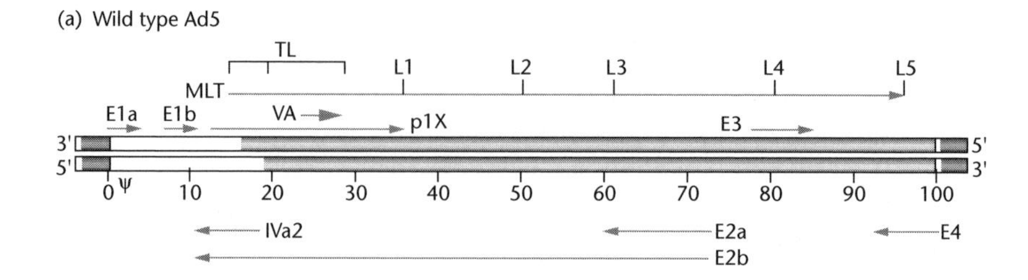

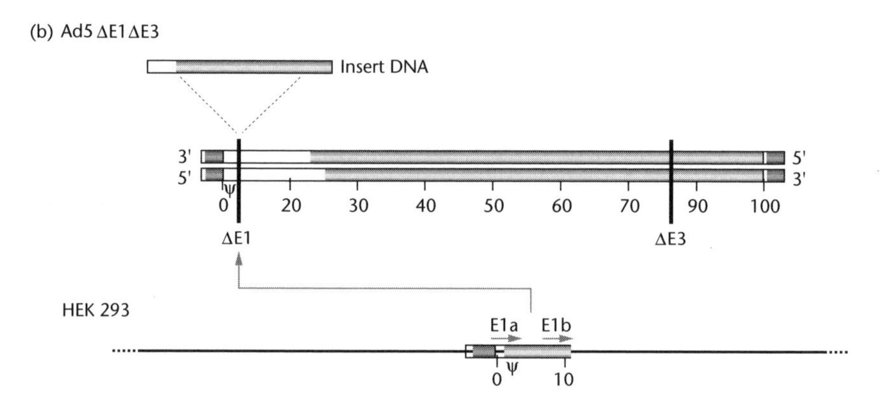

Fig. 8.3 (a) Map of the genome of wild type adenovirus Ad5. The numbers below the map divide the genome into 100 arbitrary units. The gray arrows indicate transcription units. E, early transcripts; I, intermediate transcripts; MLT, major late transcript; TL, tripartite leader, which is spliced onto the late transcripts.(b) An early adenoviral vector in which the E1 and E3 regions have been deleted to allow the incorporation of foreign DNA. The E3 region is dispensable but the E1 region is essential rendering the virus replication-deficient. The functions are supplied by the packaging line HEK 293, which is transformed with the leftmost 11% of the genome, including the E1 region.

means, e.g. by injection). The DNA is not packaged into a virus vector but is often presented in the form of a plasmid. In some cases, the plasmid is not designed to replicate in human cells and therefore the only way that long-term expression can be achieved is if the DNA integrates stably into the genome. This is a very rare event (fewer than one in a million cells are stably transformed) so the rare cells must be **selected** by including a **marker gene** in the transforming DNA that allows the cells to grow in the presence of a substance, such as an antibiotic, that is toxic to untransformed cells (Fig. 8.4). Obviously this type of strategy can only be applied to cells in culture.

For nonviral *in vivo* gene therapy, it is necessary to include the replication origin of a virus in the plasmid vector, allowing the vector to persist as an episomal replicon in the transformed cells. As discussed in Box 8.2, episomal maintenance does not continue indefinitely, and the expression of the introduced gene may therefore be short lived. To achieve *in vivo* transfer, the DNA is usually encapsulated in some form of complex that can fuse with target cells or promote uptake by endo-

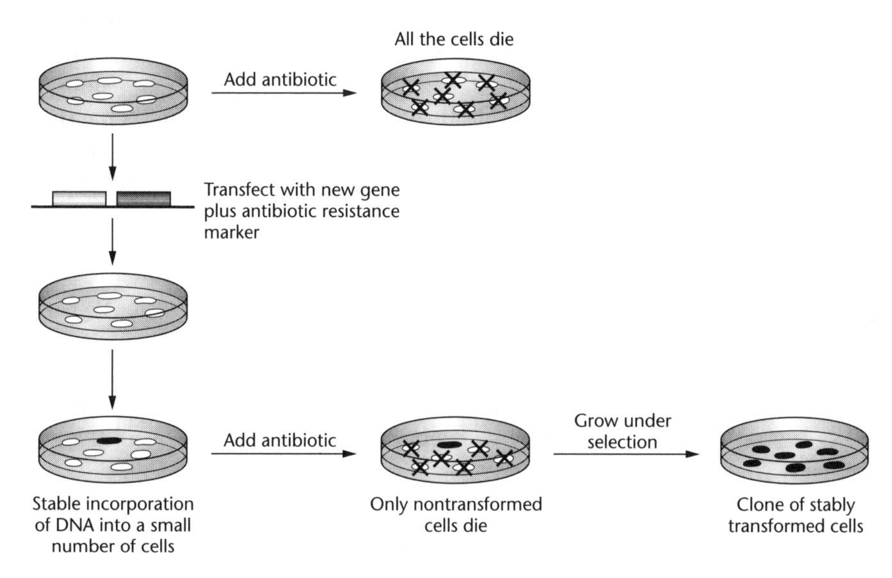

Fig. 8.4 The use of selectable markers to propagate transformed cells. If a plate of cultured human cells is exposed to the antibiotic G418, all the cells will die. However, if the cells are transfected with the selectable marker gene *neo*, which confers resistance to antibiotics such as G418, rare cells that have stably integrated the DNA will survive. This can be exploited to select cells transformed with other genes. The two genes (selectable marker and nonselected gene) are introduced at the same time and tend to cointegrate into the genome. Selection for antibiotic resistance therefore identifies cells carrying the nonselected gene too. These cells will grow under selection while cells without the new genes will die.

cytosis. For example, DNA can be enclosed within artificial lipid vesicles known as **liposomes** which are able to fuse to the plasma membrane, depositing their cargo into the cytosol. The efficiency of liposome-mediated transfection can be increased by deriving the liposome from viral envelopes, which often contain proteins that enhance membrane fusion. Such vehicles are termed **virosomes**. Unlike liposome-mediated transfection, **lipofection** involves the formation of a DNA–lipid complex (**lipoplex**) which is taken up by endocytosis. More recently, cationic polymer-based gene delivery vehicles (**polyplexes**) have become popular because specific copolymers can be used to modify the physical properties of the delivery vehicle. In some cases it has proven possible to form complexes that have different properties at different temperatures, allowing the controlled release of DNA. This could be particularly useful for *in vivo* gene transfer to particular sites in the body which could be heated or cooled as appropriate. Generally, the delivery of encapsulated DNA is much less efficient than the use of viral vectors, partly because some of the DNA is degraded before it reaches its target cells but principally because a large amount of the introduced DNA is degraded inside the cell before it reaches the nucleus. It is also more difficult to target specific cell types with nonviral transfer, although it is possible to couple the DNA to another molecule that binds specifically to receptors on particular cell types, stimulating uptake by **receptor-mediated endocytosis**.

It has also proven possible to introduce DNA directly into some tissues *in vivo*. For example, the injection of DNA solutions into muscle tissue results in the transformation of some cells, producing gene expression that may persist for a number of months. For DNA transfer to surface tissues, an alternative method is **particle bombardment**, where small metal particles are coated with DNA and accelerated into the target tissue using a high-pressure blast of air or an electrical discharge.

Case studies

Over 500 gene therapy protocols have been approved, more than half of which are designated for the treatment of cancer (see the NIH database of gene therapy trials: www4.od.nih.gov/oba/rac/clinicaltrial.htm). We discuss below several case studies that illustrate the way different types of gene therapy have been applied, and also demonstrate the risks involved. It must be emphasized that gene therapy remains a high-risk therapeutic strategy with an uncertain outcome.

In vivo *gene augmentation therapy: cystic fibrosis*

About 10% of gene therapy protocols are designated for the treatment of rare Mendelian diseases. This is because single gene defects can be addressed by the simplest correctional strategies, i.e. single gene therapy. Cystic fibrosis (CF) is the most common Mendelian disease in the Caucasian population, occurring once in every 2000 births. The defect, in the *CFTR* gene, results in the loss of a membrane-spanning chloride channel. The effects of the disease are manifest predominantly in the respiratory system and pancreas, where the chloride ion imbalance causes large amounts of thick mucus to be produced. In the lungs, this causes breathing difficulties and encourages infections, while in the pancreas, the mucus blocks the excretion of digestive enzymes leading to malnutrition. The introduction of a functioning *CFTR* gene should help to reverse these effects.

The initial CF gene therapy trials involved the use of adenovirus vectors, which naturally infect cells of the upper respiratory tract. The vectors were delivered using aerosol inhalers and in some patients, high doses of the vector caused inflammatory responses (Box 8.2). More recent trials have involved safer vectors based on adeno-associated viruses or nonviral gene delivery by liposomes. While gene transfer and limited expression have been achieved, overall results have been disappointing. A major reason for this is the inability of gene delivery vectors to efficiently penetrate the mucus that coats the lungs of CF patients.

Ex vivo *gene augmentation therapy: severe combined immunodeficiency*

The first ever gene therapy trial began in 1990 and involved a 4-year-old girl who was suffering from **severe combined immunodeficiency (SCID)**. The disease is characterized by a lack of functional lymphocytes (B cells and T cells), resulting in a complete inability to fight off infections. SCID can be caused by a number of defects but in this particular case the *ADA* gene, encoding the enzyme adenosine deaminase, was nonfunctional. The only conventional treatments available for ADA-SCID are a bone marrow transplant from a compatible donor, or regular injections with recombinant ADA enzyme. In the absence of these treatments, affected children have to live in artificial, germ-free containment facilities, and have been dubbed "bubble babies."

ADA deficiency was an ideal first target for gene therapy trials for a number of reasons:

- The pathologic effects of the disease are reversible.
- The disease results from the loss of function of a single gene.
- ADA levels vary widely in the normal population so tight control of the introduced gene is not important.
- The *ADA* gene is very small and easy to manipulate in the laboratory.
- The target cells for the therapy are lymphocytes, which are accessible, easy to culture, and easy to put back into the body of the patient.
- The alternative treatments are expensive and/or hazardous.

A functional *ADA* gene was therefore introduced into a retroviral vector and used to transduce cultured T lymphocytes, which were subsequently introduced back into the patient. Although hailed as a success at the time, the effect of the treatment was short lived and the patient continues to rely on ADA enzyme therapy to this day. Further trials were initiated in which bone marrow cells or umbilical cord blood cells were used as targets, since these populations contain the stem cells that produce lymphocytes throughout our lives. The modification of these stem cells did result in the long-term production of a small number of ADA-positive lymphocytes, but only very low levels of ADA were produced by these cells. Furthermore, ADA enzyme therapy continued throughout the trials and it is unclear whether the patients would survive without it.

In 2002 there was a major breakthrough in ADA-SCID gene therapy resulting from the use of a technique called **nonmyeloablative conditioning**, in which bone marrow in the SCID patient is partially killed in order to give the modified stem cells the chance to proliferate (Fig. 8.5). Another important factor was that none of the children in this trial had been treated with ADA. It appears that enzyme treatment may have contributed to the lack of success in previous gene therapy trials. The first patient was a 2-year-old Palestinian child who had never received ADA therapy. The new treatment seems thus far to have cured her condition and she is now living at home with her parents and producing antibodies as would any other child. She has contracted and managed to fight off chicken pox, which would almost certainly have killed her only months earlier.

Gene therapy has also been used to treat a related condition, X-linked SCID, which is caused by loss of the gamma chain of the interleukin-2 receptor. As with ADA-SCID, an *ex vivo* approach was followed in which a retroviral vector containing a replacement copy of the malfunctioning *IL2RG* gene was used to transduce cultured hematopoietic stem cells, which were subsequently reintroduced into the patient. Nine of 11 treated patients appeared to have been cured by this treatment, but since the beginning of the trial two of them have contracted leukemia, thought to be caused by the activation of an oncogene adjacent to the retroviral integration site (Fig. 8.6). For these reasons, gene therapy trials involving retroviral transduction have now been halted until further data on these patients are available.

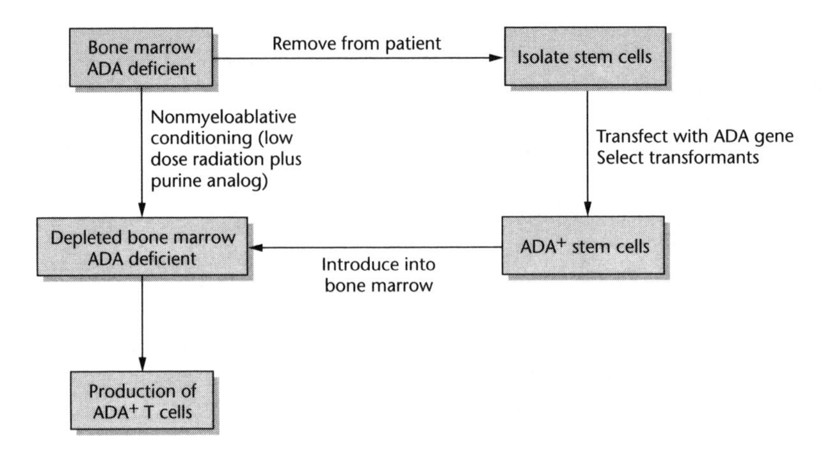

Fig. 8.5 Overview of gene therapy for ADA-SCID using nonmyeloablative conditioning.

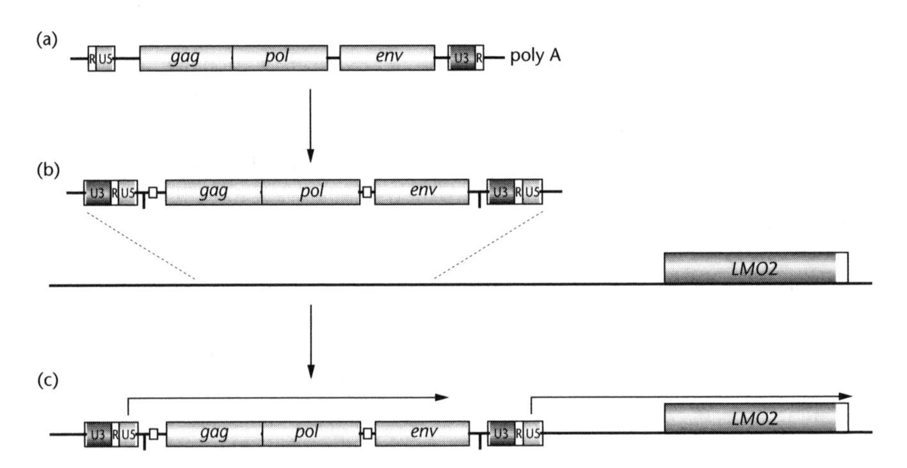

Fig. 8.6 Likely mechanism for the occurrence of leukemia in gene therapy patients treated with a recombinant retrovirus. (a) The typical structure of a retrovirus – the *gag*, *pol*, and *env* genes encode enzymes for replication and the components of the virus coat. (b) In the gene therapy vector, all the viral genes are deleted and replaced with human DNA, but the sequences required for genome integration and expression are retained. Part of the virus replication cycle involves copying the ends of the genome to generate long terminal repeats (LTRs). During this process, the LTR promoter is duplicated. (c) The duplicated promoter is able to activate adjacent genes. In two patients, clones of T cells in which the vector integrated adjacent to the *LMO2* oncogene have started to proliferate at an abnormal rate, resulting in leukemia.

In vivo *gene augmentation and inhibition therapy: cancer*

Cancer, an acquired genetic disease characterized by the excessive proliferation of cells, is discussed in detail in Chapter 5. Gene therapy is emerging as a useful alternative to conventional radiotherapy, chemotherapy, and biotherapy treatments,

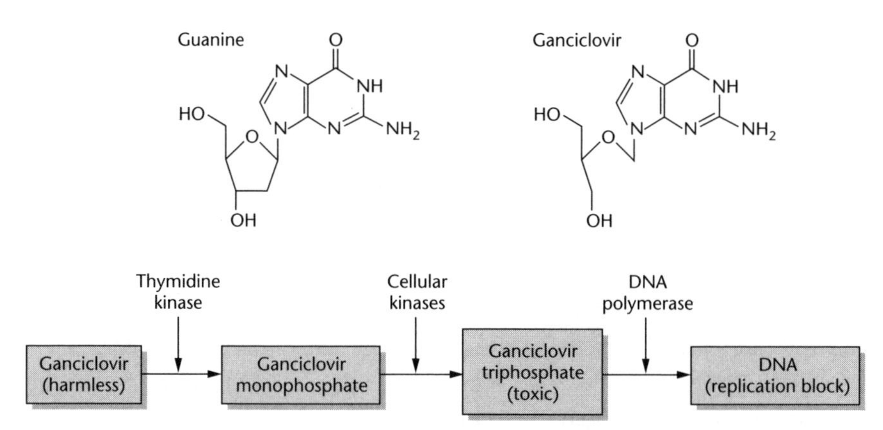

Fig. 8.7 Activity of ganciclovir. The harmless prodrug, an analog of the nucleoside guanosine, is phosphorylated by the enzyme thymidine kinase, which can be introduced into cancer cells as a suicide gene. The action of cellular kinases produces a triphosphate form which can be incorporated into DNA. DNA-containing ganciclovir is less stable than normal DNA and this stalls the DNA polymerase complex leading to a replication block until DNA repair can be carried out. If repair is impossible (as is the case when excess ganciclovir is present), the cell undergoes apoptosis.

and well over half of all approved gene therapy trial protocols have been developed for the treatment of cancer. Standard gene augmentation therapy, as discussed in the previous two case studies, can be used to replace lost tumor suppressor genes in an attempt to bring the cell cycle under normal control. However, most cancer therapy strategies involve either the inhibition of oncogenes or the killing of cancer cells. A wide variety of different strategies is available.

In the earliest cancer gene therapy trials, the tumor cells themselves were not the therapeutic target. Instead, tumor-infiltrating leukocytes were transformed with an additional copy of the gene for tumor necrosis factor, making them more efficient killers. This is an extension of the adoptive biotherapy discussed on page 127 in which T cells are stimulated *ex vivo* by the application of interleukin 2.

The use of retroviruses as *in vivo* gene therapy vectors is advantageous because these viruses can only infect proliferating cells, and are thus selective for tumor tissue. Therefore, a number of gene therapy procedures have been developed in which the tumor cells are the targets. For example, the introduction of antigen-producing genes that make the cancer cells more likely to be attacked by the immune system is a generally applicable method. Alternatively, where a specific oncogene is known to be involved, a strategy can be developed that targets the product of that gene. However, most cancer gene therapy strategies involve targeted cell killing by the introduction of suicide genes that convert a nontoxic pro-drug into a toxic derivative. An example is the thymidine kinase gene from herpes simplex virus which converts the harmless pro-drug ganciclovir into a nucleotide analog which is incorporated into DNA and blocks replication by inhibiting the DNA polymerase (Fig. 8.7).

Table 8.1 Nucleic acids used to interfere with endogenous gene expression and their mechanisms of action.

Approach	Target	Effect
Antisense oligonucleotides (DNA, RNA, or synthetic derivatives)	Gene or mRNA	Triplex formation, inhibition of transcription Duplex formation, inhibition of translation, promotes degradation
Ribozyme	mRNA	Catalytic degradation of mRNA
Maxizyme	mRNA	Catalytic degradation of mRNA with allosteric modulation
Double-stranded RNA	mRNA	Catalytic degradation of mRNA by RNA interference
Short interfering RNA	mRNA	Catalytic degradation of mRNA by RNA interference
Oligonucleotide aptamer	Protein	Inhibition of protein activity

Nucleic acids as drugs

Gene therapy involves the introduction of functional genes into human cells. The principle is that these genes are expressed, either episomally or as integrated transgenes, and the product of the gene either compensates for the original defect or counteracts it. A subtly different approach is to use nucleic acids as drugs. In this case, the nucleic acid is not an expressed gene but an exogenously applied molecule that is taken up by cells and is destined to interfere with the expression or activity of an endogenous gene at either the DNA, RNA or protein level (Table 8.1). The introduction of these drugs is achieved using the same nonviral gene transfer methods discussed above. However, unlike gene therapy, the effects of nucleic acid drugs are not expected to be permanent. They may help to cure infectious diseases or cancers, but only provide symptomatic relief for inherited disorders.

Antisense drugs

Antisense drugs are short oligonucleotides (usually 12–25 nucleotides in length) containing the complementary sequence to the coding strand of a particular target gene. They can inhibit the expression of a malfunctioning gene in two ways. The first is by interacting directly with the DNA and forming a stable triplex that prevents transcription. The second is by interacting with the mRNA and forming a stable duplex. This inhibits protein synthesis and also leads to selective degradation of the target mRNA by the enzyme RNase H, which recognizes double-stranded nucleic acids and degrades the RNA component. In the case of RNA–RNA duplexes, both strands are degraded, while only the RNA strand of DNA–RNA duplexes is degraded. Therefore, DNA oligonucleotides are more efficient therapeutic agents because they can serially interact with many transcripts. While DNA and RNA oligonucleotides are convenient to synthesize, they are prone to degradation by cellular nucleases. Therefore, today's antisense drugs are structural derivatives of natural nucleic acids (e.g. phosphorothioate-modified DNA) or synthetics that

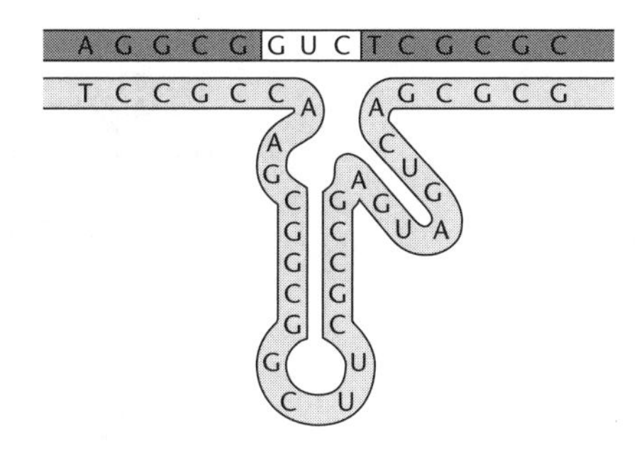

Fig. 8.8 The structure of a ribozyme. The top sequence (dark box) represents a target mRNA. The bottom sequence (lighter box) represents antisense RNA within which the catalytic component of the ribozyme is embedded. The target of the ribozyme is the boxed GUC sequence, which is cleaved resulting in the mRNA being targeted for degradation.

contain chemical bonds resistant to cellular nucleases (e.g. peptide nucleic acids, in which the bases are presented on a synthetic peptide backbone, and molecules based on nylon or carbazate backbones). Generally, the effect of any given antisense oligonucleotide is unpredictable and must be established empirically. There may also be nonspecific toxicity effects when synthetic molecules are used. Morpholino-oligonucleotides are highly efficient, nuclease-resistant molecules based on RNA which appear to address many of these problems. The first antisense drug to be approved was fomivirsen (Vitravene), a phosphorothioate-modified DNA oligonucleotide used to treat cytomegalovirus infections.

Ribozyme drugs

Ribozymes are catalytic RNA molecules, i.e. RNA molecules that have the same functional propensity as protein enzymes. **Maxizymes** are modified ribozymes whose activity can be controlled by the application of a small regulatory molecule. The natural function of many ribozymes is to catalyze the cleavage of other RNA molecules, and this activity can be harnessed and exploited to target the transcripts of specific malfunctioning genes by incorporating antisense sequences into a **ribozyme construct** (Fig. 8.8). The antisense sequences allow the ribozyme or maxizyme to become associated with the target mRNA by complementary base-pairing and the transcript is cleaved, resulting in its rapid degradation. Because ribozymes are catalytic, they have the potential to cleave many copies of the target transcript before they are degraded. Research has focused on the use of hammerhead ribozymes, which can be incorporated into antisense constructs using recombinant DNA technology. The ribozyme drug Herzyme, which targets the mRNA for epidermal growth factor 2, is discussed in Box 4.1.

The potential of short interfering RNAs

RNA interference (RNAi) is a cellular defense mechanism discovered in the nematode worm *Caenorhabditis elegans* in which double-stranded RNA can induce

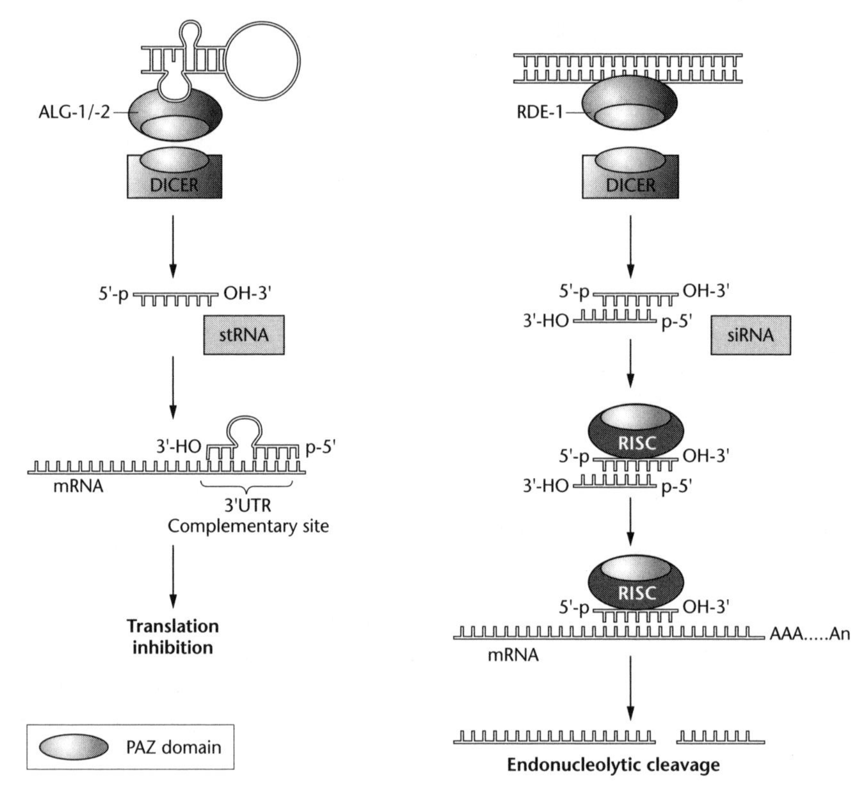

Fig. 8.9 The mechanism of RNA interference. Double-stranded RNA is recognized by a protein complex that interacts with the endonuclease DICER. If the dsRNA is imperfect and contains loops and bulges, it is processed by DICER into short single-stranded molecules that can bind to mRNA and interfere with translation stoichiometrically. However, perfect duplexes are processed into short, double-stranded molecules with short overhangs. These are known as short interfering RNAs (siRNAs) and they associate with proteins to form an RNA-induced silencing complex (RISC). The RISC catalytically degrades mRNAs carrying a matching sequence. The silencing effect is extremely potent and long lasting. There may be some cross-talk between the two pathways shown.

potent and very specific silencing of a homologous gene. The mechanism is thought to have evolved to protect cells from viruses, and involves the cleavage of dsRNA by a nuclease called **Dicer** into 21–25 bp segments known as **short interfering RNAs (siRNAs)** which are subsequently processed into single strands. These single strands associate with proteins to form an **RNA-induced silencing complex (RISC)** that recognizes complementary mRNAs and degrades them very efficiently (Fig. 8.9). Conventional RNAi does not work in human cells because the levels of Dicer endonuclease are limiting and long dsRNA molecules provoke a nonspecific interferon response against viruses. However, siRNAs can be chemically synthesized and introduced into cells (or expressed within them as "hairpin" genes) leading to the formation of RISCs and very specific silencing effects. Although there are no RNAi drugs currently available, the technique has been extensively used for

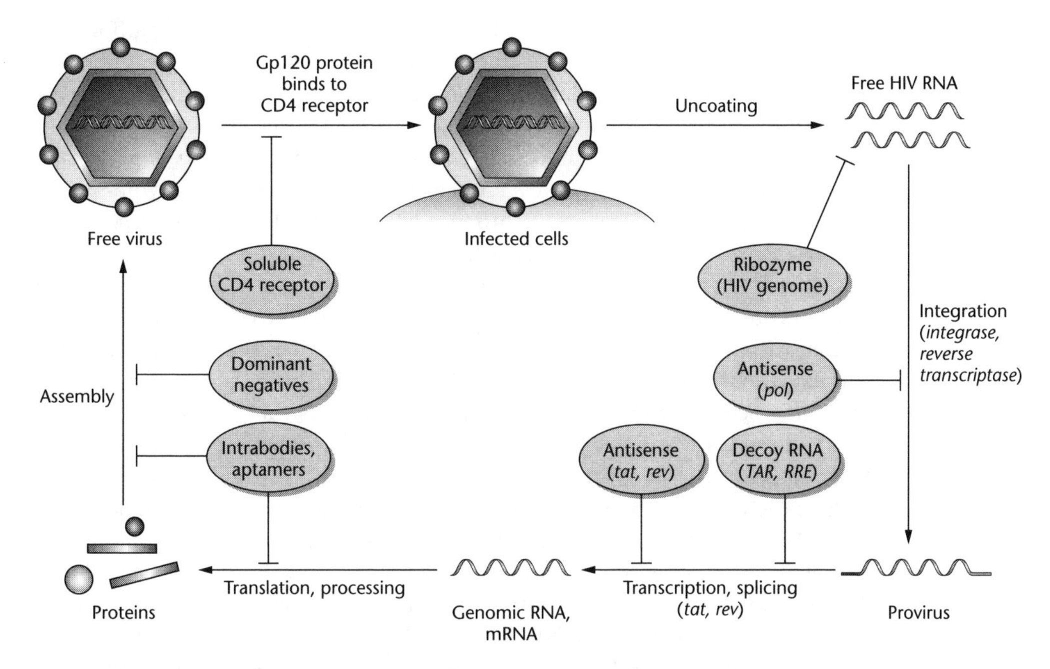

Fig. 8.10 Stages of HIV infection and examples of intervention points using gene therapy or other forms of gene medicine.

gene inactivation and functional testing in human cells (see below) and therapeutic applications are inevitable.

Aptamer drugs

The nucleic acid drugs discussed above each interact with the **transcript** of the target gene resulting in the absence of a functional protein. However, nucleic acids can also interact with proteins with great specificity and this provides another opportunity for therapeutic intervention. Oligonucleotides that interact with proteins are known as **aptamers** and can be introduced directly into the cell by conventional transfection routes. An example of the therapeutic use of aptamers is the interruption of viral replication cycles with aptamers that bind strongly to the coat protein subunits, preventing their assembly.

Gene medicine for infectious diseases: HIV

Gene therapy and nucleic acid drugs overlap to a large degree because the oligonucleotides that can be introduced directly into cells can also be expressed from small transgenes. The expression of antisense RNA, ribozyme constructs, short interfering RNAs, **intramers** (RNA aptamers expressed inside cells for protein interference), and **intrabodies** (antibodies expressed inside cells for protein interference) have all been explored as therapeutic strategies, particularly in the fight against HIV infection and AIDS (Fig. 8.10).

Human immunodeficiency virus (HIV) is the most important viral pathogen of humans and research into novel therapeutic approaches for disease prevention and treatment therefore attracts large amounts of funding. Since HIV is a retrovirus, therapeutic strategies that target either the RNA genome or the viral proteins are appropriate. HIV specifically infects T lymphocytes, making *ex vivo* gene therapy the most suitable approach. However, nucleic acid drugs that target T lymphocytes *in vivo* are also being developed. Antisense oligonucleotides have been produced that recognize the *pol* and *env* genes that encode the viral reverse transcriptase (essential for replication) and envelope proteins (essential for interaction with receptors on T cells). T cells have also been transformed with constructs expressing antisense RNA against these genes, and against those of the regulatory proteins Tat and Rev. Ribozymes have also been developed which cleave the HIV genome and these have been introduced into T cells as retroviral constructs. During the normal infection cycle, Tat and Rev bind to regulatory sequences in the genome known as *TAR* and *RRE*. A novel strategy to inhibit HIV infection is the high-level expression of **decoy RNAs** carrying multiple copies of these sequences. The rationale is that such RNAs would sequester all the Tat and Rev proteins and therefore limit their availability to the virus. Both aptamers and antibodies have been produced to block the assembly of HIV virions, and intrabodies have been expressed in T cells to inhibit the Tat and Rev proteins and block the assembly of gp160 coat protein subunits. An alternative approach to block virion assembly is to overexpress a mutated form of a viral protein that interferes with the function of the other proteins (the **dominant negative** strategy). For example, a modified Rev protein that cannot form complexes prevents the export of HIV RNA from the nucleus of infected T cells. As is the case for cancer gene therapy, further strategies have been developed with the aim of making T cells more potent killers of HIV-infected cells.

DNA vaccines

Conventional vaccination involves the presentation of antigens to the immune system either by the supply of a whole organism (killed or attenuated pathogen) or a component thereof produced by recombinant DNA methodology. **DNA vaccination** is another branch of gene medicine in which DNA is introduced into the vacinee. This does not generate an immune response **against the DNA molecule**, but if the DNA is expressed to yield a protein, that protein can stimulate the immune system to raise the appropriate antibodies (Fig. 8.11). DNA vaccination is therefore very similar to gene therapy (i.e. the introduced DNA is expressed), but instead of treating an ongoing disease directly, the disease is either treated or prevented by the activation of the immune system. The advantages of this method include its simplicity (standard cloning methods are used) and the fact that essentially the same methods can be used to vaccinate against any disease. Also, with ongoing sequencing projects for many microbial pathogens, DNA vaccination provides a rapid way to evaluate new vaccination targets using high-throughput assays in mice. The DNA may be administered by injection, using liposomes or by particle

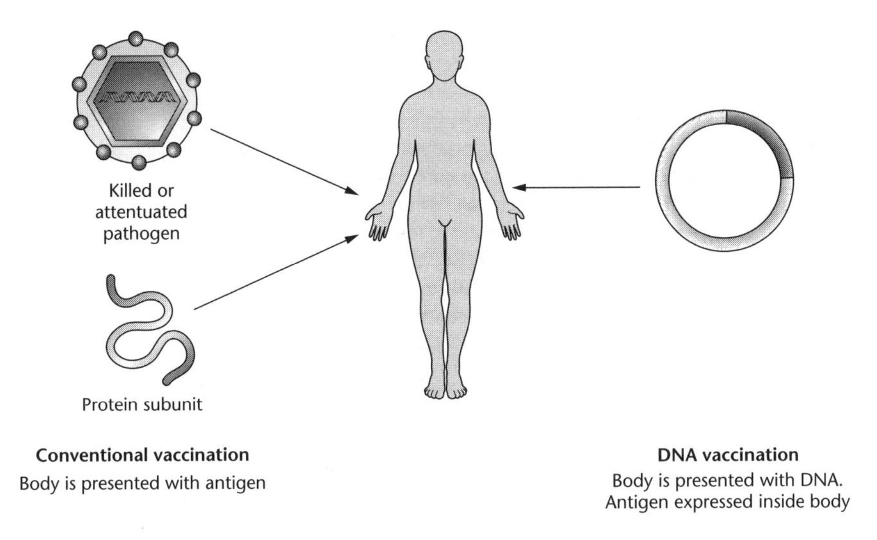

Fig. 8.11 A comparison of conventional vaccination and DNA vaccination.

bombardment. The first demonstration of the method achieved longlasting protection against influenza and currently over 40 DNA vaccination trials are in progress involving diseases such as measles, HIV, Ebola virus, and tuberculosis. DNA vaccination is also being tested as a means to provide protection against prion diseases.

Disease models

The methods of gene transfer discussed above – transfection, direct transfer, and viral transduction – can be applied to animals as well as humans. Whereas gene transfer in humans is used to prevent and treat diseases, in animals the aim is often to **cause** them, providing **disease models** that can be used to study diseases at the molecular level and test novel drugs. Cultured cells and experimental animals have been used for many years to study disease, but models could initially be obtained only by random mutation. Gene transfer technology now allows specific DNA sequences to be introduced into the genome of animal cells or the existing genome to be modified in a precise and predetermined way. Therefore, mutations can be designed and introduced into particular genes to mimic human diseases exactly. The most relevant model organism for most human diseases is the mouse, although the rat, the zebrafish and even more primitive animals such as flies and worms provide useful models for some diseases (Table 8.2).

Models of single gene disorders

Many human diseases reflect the loss of function for a single gene, and useful models can be generated by inactivating the corresponding gene (the ortholog) in

Table 8.2 Animals and microbes used to model human diseases.

Species	Properties
Mouse	The standard disease model. Mouse and human genomes have been mapped and sequenced. They show extensive synteny (conserved gene order) so gene orthologs simple to identify and genuine homologous diseases can be identified. Large scale mutagenesis screens have been carried out and specific mutants can be mapped easily using interspecific backcrosses. Gene knockout technology allows predefined mutations to be introduced into any gene (see TBASE, a transgenic/targeted mutation database: http://www.gdb.org/Dan/tbase.html)
Rat	Rats are suitable for the study of some complex diseases for which no mouse models are available (e.g. certain behavioral disorders). They are more amenable for physiologic and pharmacologic investigations
Zebrafish	A good model for vertebrate development, genetically amenable and with robust transparent embryos. Many mutant phenotypes identified in zebrafish genetic screens resemble human diseases, and particularly useful models are available for hematopoietic disorders and cardiovascular disease
Invertebrates	The invertebrate model organisms *Drosophila melanogaster* and *Caenorhabditis elegans* provide useful models for certain diseases affecting highly conserved pathways such as insulin signaling (*C. elegans*, model of diabetes) and growth factor signal transduction (*D. melanogaster*, model of cancer). Suitable for high-throughput analysis, with *C. elegans* particularly suitable for gene inactivation by RNA interference (see Chapter 2)
Microbes	Yeast and bacteria provide models for diseases based on very highly conserved molecular processes including the cell cycle (yeast, models of cancer, and Werner's syndrome) and DNA repair (*Escherichia coli*, model of genome instability in certain forms of cancer; see Chapter 5)

another organism. For this type of approach the mouse has many advantages, but in technologic terms the main benefit is the ability to perform **gene knockout** experiments in which a single gene can be selectively inactivated by mutation (Box 8.3). Many useful mouse disease models have been created in this way, including those for Lesch–Nyhan syndrome (HPRT deficiency), cystic fibrosis, β-thallasemia, and fragile-X syndrome. In other cases, loss of function effects have been produced using antisense RNA and similar strategies (Table 8.1). For example, a mouse model of diabetes has been produced by expressing a ribozyme against glucokinase mRNA specifically in pancreatic β-cells, and zebrafish models of several diseases have been produced using morpholino-antisense oligonucleotides. As such procedures become more streamlined, there is increasing overlap between the creation of disease models and the large scale mutagenesis and interference screens used in functional genomics (see Chapter 2).

Diseases caused by dominant gain of function mutations can be modeled simply by transferring an equivalent mutant version of the gene into the mouse genome by conventional gene transfer technologies (Box 8.3). For example, mouse models of Alzheimer's disease have been produced by overexpressing the amyloid precursor protein gene and a model of the premature aging disease Werner's syndrome has been produced by expressing a dominant negative version of the *WRN* gene. Useful

Box 8.3 Gene transfer to mice

The introduction of DNA into single cells is relatively straightforward. However, ensuring that the same DNA sequence appears in all cells, as is required to model a human disease in mice, is more challenging. What is required is a way to introduce DNA into the mouse germ line, so that genetically modified gametes are produced. Essentially, this is the animal equivalent of germ-line gene therapy in humans. Any animals resulting from fertilization with these gametes will have the same additional DNA sequence in every cell and are described as transgenic.

There are now several different approaches that can be used for germ-line transformation of the mouse:

- Transfection of primordial germ cells (the cells that give rise to gametes)
- Attaching DNA to the sperm head, so that it is introduced into the egg during fertilization
- Microinjection of DNA into the male pronucleus of the fertilized egg
- Infection of the mouse embryo with a recombinant retrovirus
- Transfection of embryonic stem (ES) cells, which can colonize the embryo.

Of these methods, pronuclear microinjection and the transfection of ES cells are the most widely used.

Pronuclear microinjection is as simple as it sounds. The DNA is drawn into a fine needle and injected directly into the male nucleus of the recently fertilized egg. The DNA is incorporated into the genome (usually as multiple copies) and the resulting mice are often transgenic. Occasionally, the DNA integrates after several rounds of cell division and the mouse is chimeric (transgenic in some cells only). However, if the transformed cells contribute to the germ line, the next generation of mice will be fully transgenic (Fig. B8.3a).

Embryonic stem cells are cultured cells derived from the early mouse blastocyst. When injected into a host blastocyst they will colonize the embryo and contribute to many tissues, including the germ line. One advantage of ES cells is that gene transfer is carried out in culture,

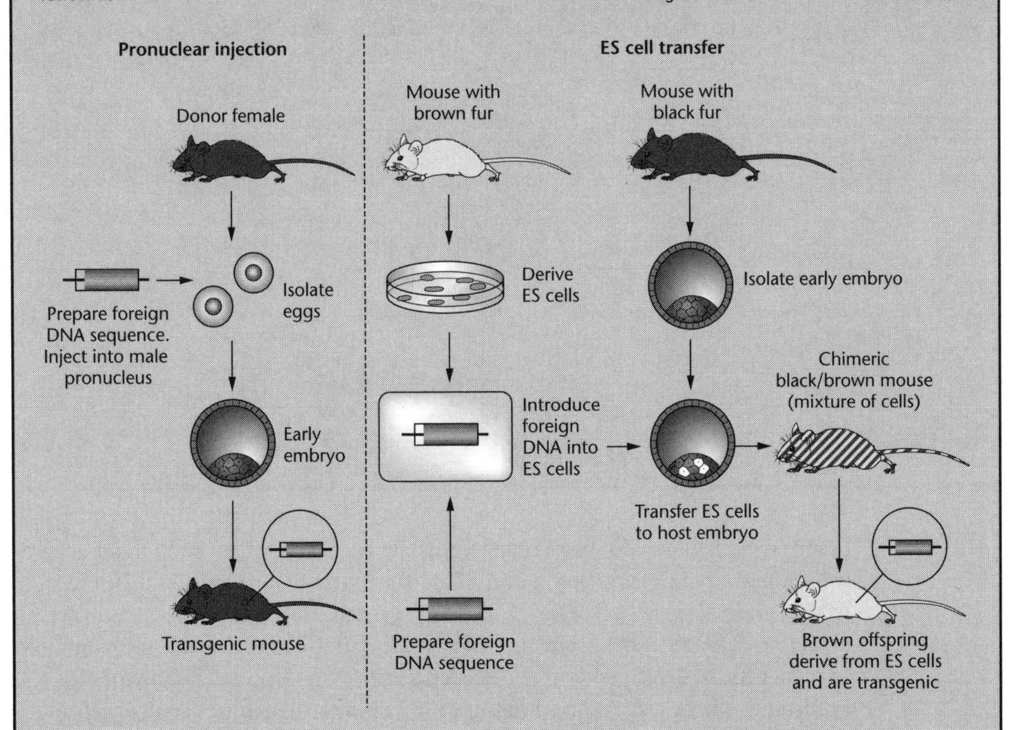

Fig. B8.3a Procedure for generating transgenic mice by pronuclear microinjection and ES cell transfer.

so selection strategies can be used to identify cells with the correct genetic modification. However, the most important benefit of ES cells is that they are amenable to recombination, which means they can be used for gene targeting. In this procedure, the introduced DNA does not integrate randomly, but undergoes homologous recombination with a related gene in the genome and replaces it. This is the basis of the gene knockout technique, which is used to create models of loss-of-function diseases (Fig. B8.3b).

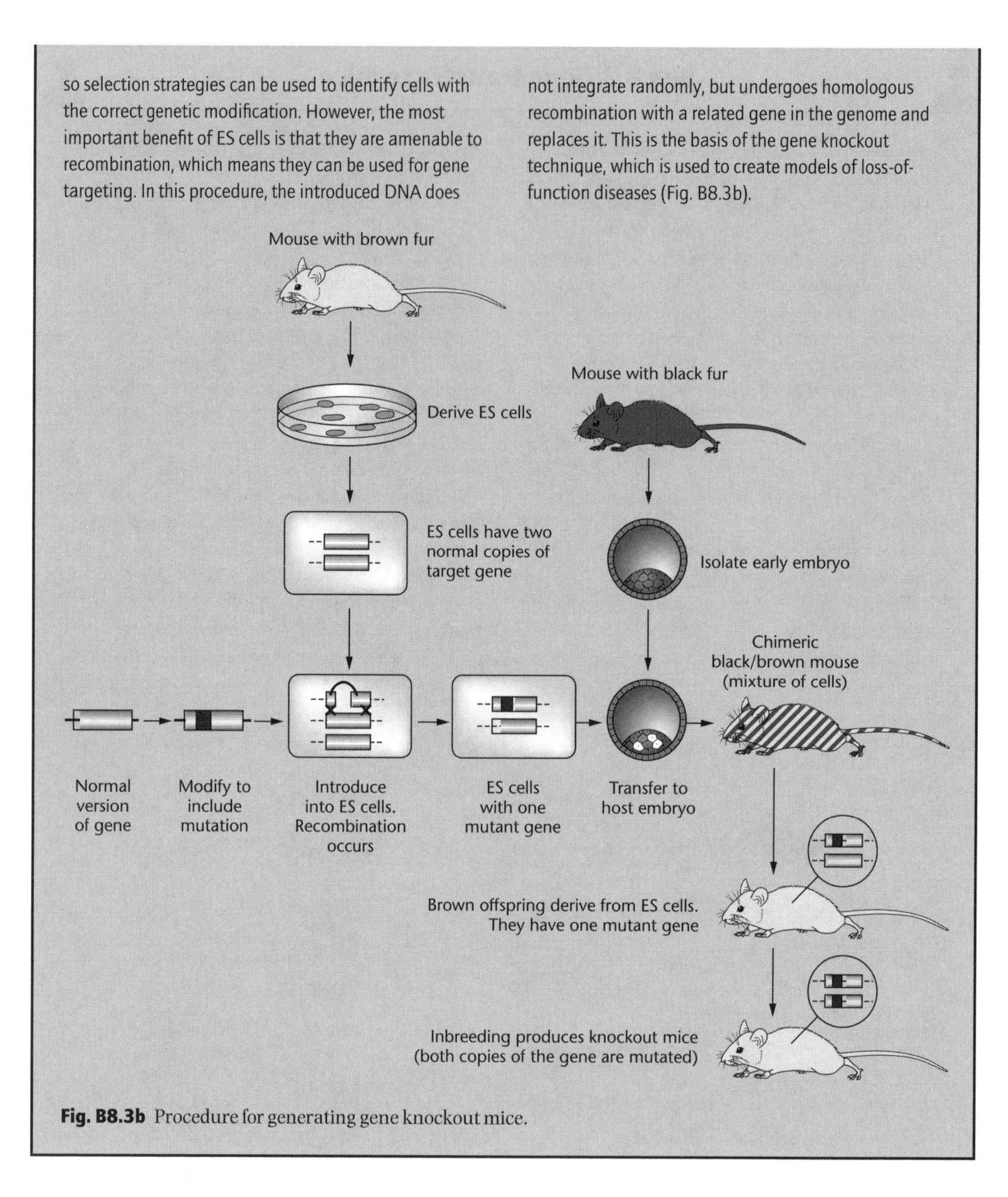

Fig. B8.3b Procedure for generating gene knockout mice.

mouse models have also been made for triplet repeat disorders such as spinocerebellar ataxia type 1, which are caused by the expansion of microsatellite repeats within genes (see p. 91). One of the first examples of gain of function disease modeling was the expression in mice of the mutant form of the prion protein to model the neurodegenerative disorder Gerstmann–Straussler–Scheinker (GSS) syndrome. There are certain limitations to disease modeling in mice, reflecting biologic and genetic differences between mice and humans. For example, the recent completion of the mouse genome sequence has shown that certain human genes have no orthologs in mice. Laboratory mice are also extensively inbred populations, while humans are outbred, so differences in the genetic background are likely to have a strong effect (see Chapter 4). Together, these contribute to a range of biochemical and physiologic differences which sometimes produce very different disease phenotypes. An example is Tay–Sachs disease, which is caused by loss of function of the *HEXA* gene encoding hexosamanidase A. This results in the accumulation of glycolipid molecules known as GM2 gangliosides in neurons of the brain, resulting in progressive neurodegeneration and eventually death. Tay–Sachs disease can be modeled in mice by inactivating the corresponding *Hexa* gene, but although GM2 gangliosides accumulate as expected, the learning difficulties and loss of motor control characteristic of the human disease do not occur.

Models of complex disorders

The availability of the human genome sequence has helped to identify genes associated with a range of complex disorders (diseases with multiple genetic and environmental components that do not show simple inheritance patterns). Modeling complex diseases is more challenging but there have been some early successes involving the crossing of mutant mouse strains to stack different mutations in the same organism. For example, offspring from a mating between *undulated* and *Patch* mice resulted in a good model for the human birth defect spina bifida occulta. In other cases, crosses like these have pointed the way to novel therapies. For example, transgenic mice overexpressing human β-globin and a mutant form of the human β-globin gene that promotes polymerization provide good models of sickle cell anemia. However, when these mice are crossed to those expressing human fetal hemoglobin in adulthood, the resulting transgenic hybrids show a remarkable reduction in disease symptoms. This suggests gene therapy using fetal hemoglobin could help to alleviate sickle cell anemia in humans.

Cell therapy

Cell therapy is the use of cells as therapeutic agents. We have touched briefly on the concept of cell therapy throughout this chapter, because in many cases the therapeutic potential of cells involves genetic modification or other forms of enhancement *ex vivo*. However, this is not always the case. Cells can be therapeutic in their own right, especially when used to replace damaged or worn out cells in the patient (a conventional example of cell therapy is skin grafting to treat burns). Much of the current excitement in cell therapy comes from the potential use of **stem cells**, which are self-renewing and therefore offer the prospect of an indefinite supply of healthy cells with prolonged therapeutic benefit.

We can distinguish between two different types of stem cells: embryonic stem (ES) cells and adult stem cells. **Embryonic stem cells** are derived from the early human embryo and have the potential to give rise to all cells types in the body, a property known as **pluripotency**. The use of mouse ES cells is discussed in Box 8.3. ES cells

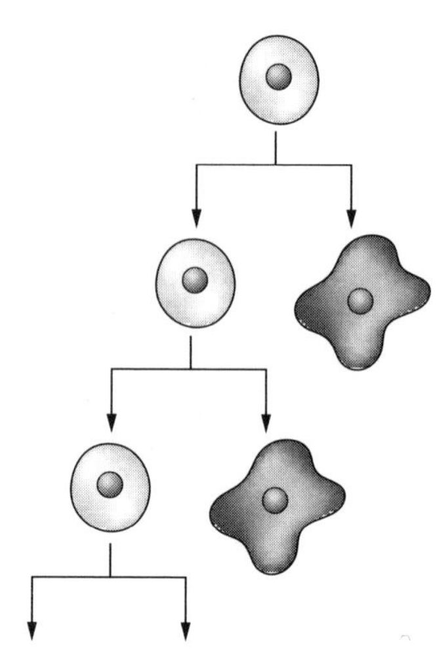

Fig. 8.12 Stem cells produce two types of daughter: another stem cell and a cell that can differentiate. In this way, stem cells can renew tissues such as skin, blood and the lining of the gut continuously throughout life. Recently, the discovery that adult stem cells can transdifferentiate (produce unexpected daughter cells, such as neurons from hematopoietic stem cells) has fueled the debate about cloning and stem cell therapy (see main text).

can be manipulated in culture to differentiate in alternative ways. In the long term, the hope is that ES cells could be used to grow replacement cells, tissues, and maybe even entire organs. Strictly speaking, ES cells are not true stem cells because they are not self-renewing at each round of cell division. True stem cells divide to produce an identical daughter stem cell and another daughter that will go on to differentiate (Fig. 8.12). Many **adult stem cells** have this property. Such cells are found in a variety of tissues and include hematopoietic stem cells (which produce all the cells of the blood and immune system), brain stem cells (which give rise to neurons and glial cells), muscle stem cells (which give rise to mature muscle cells), and stem cells in the skin. Some of these cells are usually quiescent whereas others are active throughout life (skin stem cells being a good example). Stem cells are multipotent because they can give rise to a range of different cell types, albeit not the diversity offered by ES cells. Recently, however, it has been shown that adult stem cells can in some cases **transdifferentiate**, i.e. give rise to uncharacteristic daughter cells. For example, brain stem cells have been shown to produce lymphocytes and cells of the myeloid lineage, while hematopoietic stem cells have been shown to produce neurons. This raises the exciting possibility that easily isolated stem cells, such as those in bone marrow, could be used to generate new cell populations in the brain, e.g. for the treatment of Parkinson's disease.

Stem cells and cloning

One problem with cell therapy is **hyperacute rejection**, where the immune system of the recipient recognizes the cells as foreign and attempts to destroy them.

This issue does not arise if the patient's own stem cells are used, but in many cases such cells are unavailable. One way to address this issue is by **cloning**, i.e. using nuclei from the patient's cell to replace the nucleus in a fertilized egg. The resulting embryo would therefore carry the same genetic material as the donor, and would represent a rich source of compatible stem cells. This process, known as therapeutic cloning, raises strong ethical objections not least because for some it represents the first step towards reproductive cloning (Box 8.4).

Box 8.4 The ethics of human cloning

The first mammal to be cloned from an adult cell was Dolly the sheep, created by Ian Wilmut and colleagues at the Roslin Institute in Edinburgh, UK. The procedure was simple in concept but very difficult in practice (Fig. B8.4). Nuclear transfer was accomplished by fusing somatic cells from a cultured mammalian epithelial cell line to enucleated oocytes. The donor cells were deprived of serum beforehand therefore causing them to withdraw from the cell cycle and become quiescent. This step was found to be essential for the success of the procedure. Even so, only 29 out of a total of 434 oocytes developed to the transferable stage and only one, Dolly, survived.

The birth of Dolly caused a frenzy of media interest and public debate about the possibility of human cloning and the ethical problems it would pose.

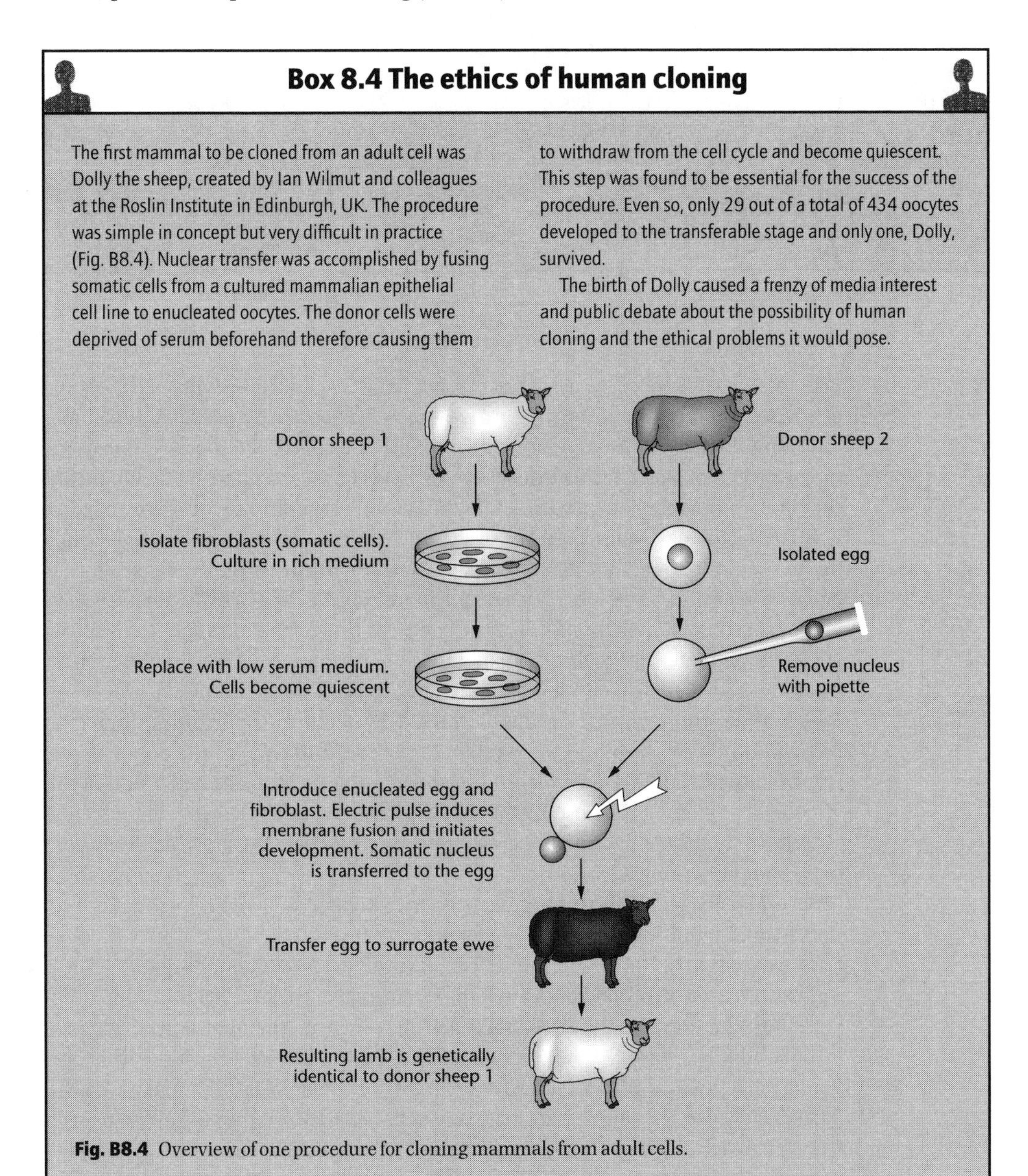

Fig. B8.4 Overview of one procedure for cloning mammals from adult cells.

It is necessary to distinguish between **therapeutic cloning**, which would be used to produce cells and tissues matching a particular patient, and **reproductive cloning**, which would produce new human individuals equivalent to Dolly.

As for germ-line gene therapy (Box 8.1), one of the major ethical objections to reproductive cloning is that the technology is imperfect and could have unpredictable effects. Indeed, since Dolly, hundreds of cloned mammals have been produced in a variety of species and many of those surviving to term have shown evidence of congenital birth defects and other diseases. Others object to reproductive cloning as a point of principle because it is an unnatural process for creating life, as opposed to IVF (*in vitro* fertilization) treatments which simply assist the normal and natural process of fertilization. Despite these reservations, one religious group in America claimed early in 2003 to have cloned at least five humans, although whether this turns out to be true remains to be seen.

Although therapeutic cloning should not incite the same ethical dilemmas as reproductive cloning, many people are still offended by the idea that a human life could be created to be exploited for "spare parts." To some, this argument is misconceived because only embryos already created in IVF clinics would be made available for cloning procedures, while others regard therapeutic cloning suspiciously and see it as a slippery slope towards reproductive cloning. At the current time, legislation for cloning varies in different countries around the world and continues to evolve in response to public debate.

Organ transplants

We close this chapter with a brief discussion about organ transplants, a traditional cell-based therapy for organ failure which is belabored by long waiting times reflecting the lack of organ donors. As with other forms of cell therapy, one of the major problems with organ transplants is hyperacute rejection, and this means that donors must be found with a matching tissue type. In the future, transgenic animals, particularly pigs, could be used to supply functional organs to replace failing human ones – an approach termed **xenotransplantation**. There is vigorous debate concerning the ethics of xenotransplantation, particularly the safety aspects of the procedure. For example, there are concerns that endogenous pig retroviruses could be activated following transplantation, perhaps even recombining with human retroviruses to produce potent new hybrids with unknown properties. Ethics aside, the technique remains limited by technical difficulties, including recognition by the host immune system of porcine antigens on the donor tissue. Hyperacute rejection is dependent on antibodies raised against the foreign organ and the activation of the host complement system. In both cases, the major trigger for rejection appears to be a disaccharide group (Gal-α-(1,3)-Gal) which is present in pigs but not primates.

Several transgenic strategies have been investigated to avoid hyperacute rejection, including the following:

- the expression of complement-inactivating proteins on the donor organ;
- the expression of antibodies against the immunogenic disaccharide group;
- the introduction of genes encoding other carbohydrate-metabolizing enzymes;
- the inhibition of α-(1,3)galactosyltransferase (the enzyme that forms this particular carbohydrate linkage, an enzyme that is present in pigs but not in primates).

In the latter case, the simplest strategy would be to knock out the α-(1,3)galactosyltransferase gene by homologous recombination. This was reported by two independent research groups in 2002. In both cases, several live piglets were born following nuclear transfer to enucleated oocytes. The donor cells were fetal fibroblasts in which the α-(1,3)galactosyltransferase gene had been disrupted by targeting with a homologous construct. One of the groups reported the birth of five healthy piglets on Christmas Day 2001 (these were appropriately named Noel, Angel, Star, Joy, and Mary) while the other claim their four piglets were born about 3 months earlier. The pigs used by the second group were of a miniature breed, which may have advantages, in terms of organ size, for human recipients. Although hyper-acute rejection has been overcome with this strategy, there may be further problems including delayed immune rejection (involving natural killer cells and macrophages) and the requirement for T-cell tolerance.

Further reading

POGM: Chapter 10 covers gene transfer to animal cells and Chapter 11 covers the genetic modification of animals, including cloning. Chapter 13 considers some of the more advanced strategies discussed in this chapter, including antisense, ribozymes, and RNA interference. Chapter 14 includes sections on gene therapy, DNA vaccines, and animal models of disease.

POGA: Chapter 10 discusses the use of large scale mutagenesis and interference screens to identify appropriate disease models.

Daley GQ (2002) Prospects for stem cell therapeutics: myths and medicines. *Curr Opin Genet Dev* **12**, 607–613.
Gordon JW (1999) Genetic enhancement in humans. *Science* **283**, 2023–2024.
Johnson M (1998) Cloning humans? *Bioessays* **19**, 737–739.
Weissman IL (2002) Stem cells – scientific, medical and political issues. *New Engl J Med* **346**, 1576–1579.

Four stimulating articles that look at the technical and ethical issues involved in cloning and stem cell therapy.

Davies JC, Geddes DM, Alton EWFW (2001) Gene therapy for cystic fibrosis. *J Gene Med* **3**, 409–417.
Kay MA, Glorioso JC, Naldini L (2001) Viral vectors for gene therapy: the art of turning infectious agents into vehicles of therapeutics. *Nature Med* **7**, 33–40.
Somia N, Verma IM (2000) Gene therapy: trials and tribulations. *Nat Rev Genetics* **1**, 91–99.

Three useful articles covering the technology, applications and risks of gene therapy.

Dooley K, Zon LI (2000) Zebrafish: a model system for the study of human disease. *Current Opin Genet Dev* **10**, 252–256.
Muller U (1999) Ten years of gene targeting: targeted mouse mutants, from vector design to phenotype analysis. *Mech Dev* **82**, 3–21.
Shastry BS (1998) Gene disruption in mice: models of development and disease. *Mol Cell. Biochem* **181**, 163–179.

These three articles describe how mice and zebrafish can be used to model human diseases.

Reyes-Sandoval A, Ertl HC (2001) DNA vaccines. *Curr Mol Med* **1**, 217–243.
A good summary of the principles and applications of DNA vaccines.

Tuschl T, Borkhardt A (2002) Small interfering RNAs: a revolutionary tool for the analysis of gene function and gene therapy. *Mol Intervent* **2**, 158–167.
A look at the current use of RNA interference as a tool for functional analysis of human genes, and at its potential as a therapeutic modality in the future.

Index

Page numbers in *italics* refer to figures; those in **bold** to tables and boxes. Index entries are arranged in letter-by-letter alphabetical order.